The Weather of Britain

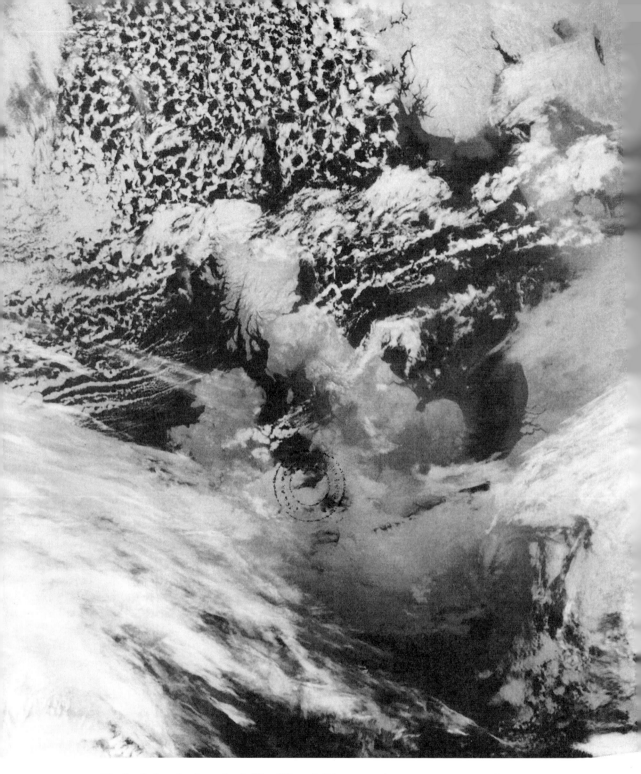

The day before the Great South-West Blizzard, February 1978,
in Devon, Somerset, Dorset and South Wales

The Weather of Britain

ROBIN STIRLING

BA Hons FRGS FRMetS

second, revised and expanded edition

dlm

First published in 1997
by Giles de la Mare Publishers Limited
3 Queen Square, London WC1N 3AU
in this second, revised and expanded edition
First published in 1982
by Faber and Faber Limited

Printed in Great Britain by
Butler & Tanner Limited

A CIP record of this book is available
from the British Library

ISBN 1–900357–06–2

Foreword

When I completed my initial forecasting course over thirty years ago I remember thinking, after six gruelling months, 'That can't be it. When do they take me aside and tell me the *secret* of weather forecasting and make me vow to use my new powers only for the good of mankind!' My head was stuffed with formulae but I had no clue about the weather. How I wish someone had given me Robin's book then. Because he knew the secret and had that almost indefinable 'feel' for the scale and process of weather, something that's taken me all my life even to approach. And, you lucky people, you too can acquire his insight just by reading this book—and keeping it—and dipping into its treasury of weather records and stories, and there are hundreds of them! Lots of recent weather books have had an American bias. Not this one. Resolutely British with only brief and almost pitying glances at our neighbours, the citizen needs no other volume to possess an authoritative account of every aspect of Britain's weather, past, present and future. I can think of no topic in meteorology which he has not covered happily and well.

The secret lies in his use of numbers. Now don't run away. I didn't say mathematics. I agree the dog-breath of maths can be a turn-off. But numbers are fine and beautiful things without which it's impossible to grasp how big, how fast and how good our weather—the best in the world—really is. As well as numbers Robin explains just how important good records are. I've always been a bit of a patriot. Britain can do no wrong, I thought. I shall never forget my distress and shame when the pictures appeared of rusting tractors in Africa in 1946 when 'our' groundnut scheme failed dismally, largely because our rainfall records were not good enough to calculate the likely variability of the local rainfall. I learned about Santa at the same time. Thank goodness I later discovered women!

This book is full of treasures. A noble defence of Manchester. A Surrey valley with as much frost as the Scottish Highlands. The drought of 1893, which puts our hose-pipe bans into perspective, with parts of London dry for seventy-three consecutive days, never equalled. And even the TORRO Hailstorm Intensity Scale. Yes, anoraks too will be thrilled! Robin is no dry academic, but my hero. I wish we had met. If the publisher wants the review copy back—he'll have to sue. Enjoy!

Ian McCaskill
BBC, London, April 1997

Contents

Preface to the Second Edition (1997)

The first edition of this book had forty-six tables, which in the eyes of some made it an academic work. It is not an academic book, however, as the reviews have shown. It is written for anyone with an interest in our weather from the age of nine to ninety. There is plenty of description so that the tables can be ignored by the reader, if desired; but they contain data which even people who are not statistically minded will be tempted to scrutinise in order to track down the snowiest winter or wettest place. An interesting addition to one of the tables considers the behaviour of insects in relation to the Beaufort Wind Scale.

Since the first edition was published in 1982 there does seem to have been a lot of weather worth writing about. Or has weather just been better publicised by the media? In January 1982 we saw record low temperatures of −26°C (−15°F) in Shropshire; in August 1990 record high readings, over blood heat, at Cheltenham. In January 1987 a cold blast of air gave maxima as low as −8°C (18°F) with sunshine and a penetrating wind in the south of England. February 1991 saw another very cold spell and with it British Rail's 'wrong kind of snow', which stopped the new rolling-stock designed for wet snow. The weather had also interrupted the trains in October 1987, but BR could not be blamed then, for that was when we had the Great Storm or 'Hurricane' in the south-east, which blew many big trees onto the track. People in country places had to eat up the contents of their freezers because of electricity failure. February 1990 saw a storm as severe as that had been, which affected London, and was far more widespread and caused more loss of life as it occurred in daylight. Tidal surges affected North Wales and devastated many homes. Another storm hit the headlines when the tanker *Braer* went aground off the Shetlands in January 1993. Over much of England there appears to have been a rash of good summers. 1983 had the warmest July on record, 1995 the warmest August. November 1994 achieved a record for mildness, and October 1995 saw remarkable warmth. From 1982 to 1995 there have been only two significant snowfalls in Surrey. Since 1853 in the Bristol area eight out of ten of the warmest years have occurred since the first edition came out. Is global warming a reality? The thirty-year averages don't support this, but that is only up to 1990. Four summers in the 1990s—1990, 1992, 1994 and 1995—would tip the averages if they were within the standard thirty-year period. The winters were mild, too.

Yellow rather than green grass has been a common feature of the past fourteen years. Water shortages have been all too common, Yorkshire appearing to be badly affected. Genuine lack of rain and media excitement about leaks and excessive pay have brought the issue of drought before the public eye.

In general, all the tables of statistics have been updated where possible. Unfortunately information on rainfall duration is not readily available, and so existing data has been used, supplemented by information from some amateur stations. But the coverage is not what I would myself have chosen as the stations are not located where most of us live. The table showing days over 25°C (77°F) has gaps. Alas, one of the by-products of slimming down the armed forces is that some long-period meteorological stations have gone out of service. The London area, where many people live and where readership of this book should be at its highest, lacks long-period comparable records. Heathrow didn't begin keeping records till the 1940s and Kew Observatory with its hundred years of records closed in 1980. After that point data for Kew Gardens has been used instead. The professional meteorologist may not be too happy with this, but as far as the general reader is concerned it makes little difference.

The county boundary changes of April 1996 were brought into force after much of the new edition of the book had been prepared. They have not, therefore, been included.

Preface to the First Edition (1982)

I have written this book because I believe that there are many people in Britain who would like to know more about our weather. A substantial proportion of the population listen to, or look at, the forecast on the radio or on television, or in the newspapers.

Ever since I can remember, I have always been interested in weather. Perhaps the seeds of interest were sown by some stranger. Or was it the view I had from my bedroom when staying at a cottage in Somerset? From there, early in the morning, I could see the mist rising and falling in the valley, enveloping the cows and trees and the roofs of Crewkerne one moment, and revealing the early morning milk train on the Southern Railway the next.

My first instrument was an ordinary thermometer, a Christmas present. My family probably wished they had not given it to me, for it turned out to be a very cold, frosty Yuletide, and frequent excursions to the garden were not viewed with favour. With the minimum temperature well below –7°C (20°F), bitter draughts swept through the house whenever an outside door was opened.

Some people's weather enthusiasm is rainfall; for others it is thunder or hard frost for skating, or deep snow for tobogganing. These are all described in the book, as well as the rarer events such as tornadoes and waterspouts, freezing rain and golfball-sized hail. Cricket and tennis fans, sailing enthusiasts, walkers and the vast majority of us who are always travelling for some reason will find in it plenty to stimulate their interest, just as my interest was stimulated by A.B. Tinn's *This Weather of Ours*.

Acknowledgements

On behalf of my late husband, Robin Stirling, I should like to thank the staff of the National Meteorological Library for their help. Particular thanks are also due to Kevin Bromley, Cyril Ward and Chris James for their painstaking proof-reading and helpful advice, and for their assistance with the implementation of corrections; to John Pritchard and Gillian Watts for checking the second draft of the manuscript and reading the proofs; and to Ian Currie and Storm Dunlop for their helpful advice.

Before my husband's death in June 1996 he asked for two chapters to be examined by meteorological specialists. Terence Meaden of the Tornado and Storm Research Organisation kindly performed that service in respect of the chapter on twisters, whilst Terry Marsh of the Institute of Hydrology substantially enhanced the chapter on drought. To these professionals I would like to express my gratitude.

Finally, my biggest thank you is to Michael Herbert, who steered Robin through the intricacies of WordPerfect 6, and has given freely of his time and expertise in preparing charts, producing drafts and inserting corrections. Without him the book could not have been published.

Sheila Stirling

Illustration Credits

Abbreviations

B.R.	*British Rainfall,* first published 1860. Since 1969 renamed *Monthly and Annual Totals for the United Kingdom.*
CET	*Central England Temperature series.* A composite record of monthly temperatures since 1659, begun by the late Professor Gordon Manley and now updated each month. Details appear in *Weather* magazine four times a year. The original data appeared in Q.J.100, pp. 389–405.
C.O.L.	*Climatological Observers' Link.* Terry Mayes, 9 Grayling Drive, Colchester, Essex CO4 3EN. Gives information about weather at amateur stations. Publishes twelve monthly summaries, usually within four weeks of the end of the month, and an annual summary. First published 1970.
D.W.R.	*Daily Weather Report,* Meteorological Office, Bracknell, Published 1860–1980.
J.M.	*Journal of Meteorology,* Dr. G.T. Meaden, 54 Frome Road, Bradford-on-Avon, Wilts. First published 1975. Ten issues per year. General-interest articles and reports of tornadoes and thunderstorms in Britain.
L.W.C.	London Weather Centre.
M.M.	*Meteorological Magazine,* Bracknell. First published 1866. Ceased publication 1993.
M.O.	Meteorological Office, London Road, Bracknell, Berkshire RG12 2SZ (not to be confused with Met.O. which refers to a series of M.O. documents).
M.W.R.	*Monthly Weather Report.* First published 1884. Summarises month's weather at Meteorological Office stations, i.e. those inspected and approved by the M.O.
Q.J.	*Quarterly Journal* of the Royal Meteorological Society. Contains research articles. Reading, Berks. First published 1871.
TORRO	Tornado and Storm Research Organisation, Oxford Brookes University, OX3 0BP.
W.	*Weather* magazine (monthly). Royal Meteorological Society, Reading, Berks. First published 1946. Articles of general interest, some research articles and daily synoptic charts for noon. Monthly summary for selected British and European stations.
W.M.O.	World Meteorological Organisation, Geneva.

Weather and Wind Symbols and Conversion Tables

Temperature conversion
°C   °F

WEATHER

○ clear sky	◕ ¾ cloud	
◔ ⅛ cloud	◕ ⅞ cloud	
◑ ¼ cloud	● overcast	
◑ ⅜ cloud	●• rain	▽ rain shower
◑ ½ cloud	●✳ snow	✳ snow shower
◓ ⅝ cloud	⇕ hail	, drizzle
⊗ sky obscured	≡ mist	≡ fog

sleet is shown by a rain and snow symbol

warm front
occlusion front
cold front

WIND DIRECTION AND STRENGTH

◎ calm	
○—	force 1
○⌐	force 2
○⌐	force 3
○|||||	force 8
○◀	force 10

inches	mb
31·0	1050
	1040
30·5	1030
	1020
30	1010
29·5	1000
	990
29	980
	970
28·5	960
	950
28	

°C	°F
40	104
35	95
30	86
25	77
20	68
15	59
10	50
5	41
0	32
-5	23
-10	14
-15	5
-20	-4
-25	-13
-30	-22

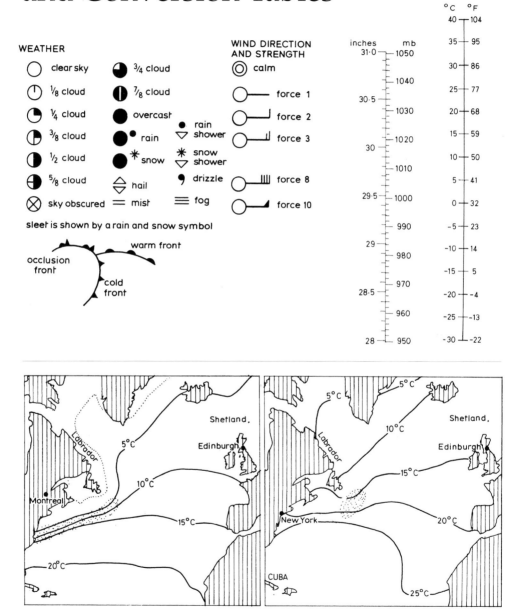

Fig.1. (*left*) **Average North Atlantic sea-surface temperature in January**

Fig. 2. (*right*) **Average North Atlantic sea-surface temperature in August**

1 | The Weather and the Sea Around Us

> The sea temperature is thus one of the masters of the weather.
> But it is also a slave, because the weather affects the sea temperature.
>
> Nigel Calder, *The Weather Machine* [1]

Wind brings weather. Weather brings wind. And from whatever direction the wind blows it must cross the sea to reach Britain. That may seem a very obvious fact but it is one which it is vital to appreciate when studying our weather. Fig. 3 shows the names given to the seas around us, as mentioned in the Shipping Forecast, broadcast by BBC Radio 4.

The maps in Figs. 1 and 2 give some important facts about the sea-surface temperatures for January and August. What stands out clearly is the relative warmth of the water around the British Isles. The reason for it is the apparently endless supply of warm sea-water of tropical origin that washes our shores. The warm water is carried there by the combined forces of the Gulf Stream, or, as it is called nowadays, the North Atlantic Drift, and the prevailing south-west wind. Most of us realise how fortunate this is for us when we compare our weather with that commonly experienced in the same latitudes on the other side of the Atlantic in Newfoundland and Labrador. There, the iceberg-laden Labrador current affects the coastline so that winds from the sea bring cold, raw days in summer and much snow in winter. The climate is so bleak that the northern limit of tree growth is reached in mid-Labrador, lying about the same distance from the equator as Edinburgh where some sub-tropical species grow well in the Royal Botanic Gardens.

The sea-surface temperature around the British Isles changes comparatively little between summer and winter. Off Cornwall, for example, the range is only 6°C (11°F), from 10°C (50°F) to 16°C (61°F). By contrast, the land nearby is several degrees cooler in winter and several degrees warmer in summer. Yet both surfaces receive the same amount of heat from the sun. On land, the sun's heat warms only a shallow depth of ground, with the result that the warmth is readily available to warm the air in contact with it but the heat is also easily lost or radiated away at night and during the winter. At sea, the same amount of heat is spread much further down into the depths. In consequence, the sea-surface temperature changes are usually rather slow because of the high thermal capacity of the water, or its ability to retain heat. The sea heats and cools slowly, the land much more rapidly. This has important results for air in contact with either of those surfaces. Air in contact with the sea for any length of time will tend to take on the temperature of the sea, and so winds from the Atlantic will

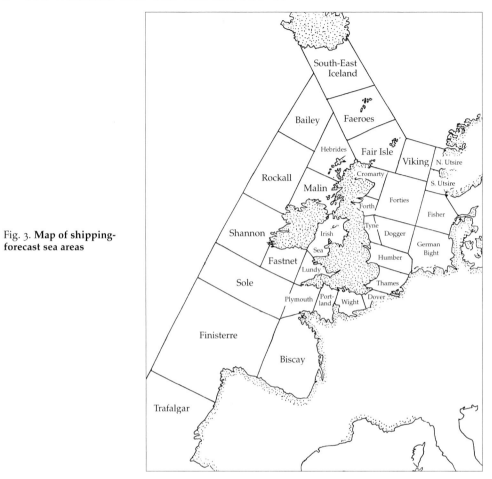

Fig. 3. **Map of shipping-forecast sea areas**

be mild in winter and cool in summer. Winds from the landmass of Europe will tend to be cold in winter and warm in summer. But, while over the sea, the air will also absorb much water vapour essential for rain.

It seems, then, that those who profess to dislike our relatively damp and cloudy summers can pin some of the blame on the ocean. However, the relatively warm sea is really a good thing. If the sea temperatures were lower, we should, most likely, have less rain and would thus be short of water: for there would be less evaporation, less moisture in the air, and so less to yield as rain. We can also work outdoors almost every day of the year, digging the garden, ploughing the land, pruning fruit-trees, and building houses and roads. Even concreting can often be done in midwinter. It is possible to play football, rugger and hockey in the coldest months. Both golf and tennis have their substantial winter following, too. All these outdoor pleasures are denied to many others living in our latitudes, like Canadians for example, because of the cold and snow-covered ground. The same goes for many Germans.

In January, the warmest sea-water, with a temperature of around 10°C (50°F), is found off south-west Ireland and Cornwall; the coldest occurs, not to the north of Scotland, as might be expected, but off the coast of Norfolk. The reason for this, too, is the influence of the North Atlantic Drift which, as we have seen, carries sub-tropical water north-eastwards, passing to the north of Scotland and on to the north of Norway.

The winds which blow over the British Isles come from three main sources: the Atlantic, which is commonest, the Arctic and Europe. When the winds blow from the Atlantic, the temperature of the air will not be very different from that of the sea surface over which the air has moved, and so it will be mild in winter and cool in summer. In fact, the general weather over Britain, particularly the west coast and the islands, will be like that found over the ocean. Out in the Atlantic, at Weathership 'J', rain is likely on about 300 days a year and temperatures below freezing point only occur once or twice each winter during incursions of cold air from the Arctic.[2] When Atlantic weather prevailed over the country in December 1974, January 1975, and November 1994, for example, there was no frost at all in many parts.

However, an east wind can bring severe weather for a few days in most winters and occasionally for longer, as occurred between 22 January and 16 March 1947, and again between 23 December 1962 and the beginning of March 1963, when most of eastern Britain was snow-covered. The surrounding warm defence of sea-water is weakest where the waters are narrow, off Kent and East Anglia, and cold winds crossing the sea will be least warmed there. Table 1 shows noon temperatures at Heathrow and in the Netherlands during the remarkably severe February of 1956, when east winds blew on most days, bringing temperatures some 5°C (9°F) below average to many parts of England.[3] It will be noted that on almost every day Heathrow was noticeably less cold than De Bilt, the short sea crossing for the cold wind being sufficient to warm it up by quite a few degrees, if not enough to prevent London from experiencing a remarkably bitter spell. Table 1 also gives data for a shorter but more recent cold spell in 1987 to show the increased severity of the cold on the other side of the North Sea. The noon temperature in Holland was –7°C (19°F) or lower for a whole week, a very severe frost. It should also be noted that the cold began a day or so earlier than in the south of England and lasted a day or two longer.

It is also normal in wintry spells brought by easterly winds for the temperatures to be lowest in the south. From Lincolnshire northwards, the extra width of the North Sea permits greater warming as the cold wind is in contact with the relatively warm sea for longer; but sometimes this extra warming is a mixed blessing since increased cloud and rawness make the conditions seem more unpleasant and feel colder.

If the North Sea, 600 kilometres (370 miles) wide from Yorkshire to Denmark, has a significant effect on cold winter winds, the influence of 800 kilometres (500 miles) of sea which is a few degrees warmer, between Scotland and Iceland, is even more dramatic. That is the reason why the north wind is not as cold as might be expected, because it

Table 1 **Noon Temperatures at Heathrow and in The Netherlands in Two Severe Cold Spells**

Date 1956	DE BILT (°C)	HEATHROW (°C)	Date 1987	DEN HELDER (°C)	HEATHROW (°C)
January 30	−3	8	January 9	2	2
31	−7	−1	10	−6	0
February 1	−12	−4	11	−9	−3
2	−6	−4	12	−7	−7
3	1	−1	13	−7	−5
4	−1	0	14	−12	−3
5	−1	7	15	−7	0
6	−1	5	16	−7	1
7	−3	5	17	−6	−2
8	2	6	18	−6	−2
9	−6	1	19	−5	−2
10	−5	−1	20	−3	2
11	−6	−2	21	0	6
12	−4	3	22	3	7
13	−6	2			
14	−4	−1			
15	−8	2			
16	−11	2			
17	−8	1			
18	−5	−1			
19	−5	−1			
20	−6	−1			
21	−5	−1			
22	−12	−1			
23	−8	0			
24	−7	−2			
25	−2	1			
26	−1	4			
27	2	2			
28	−1	7			
29	7	11			

De Bilt lies 400 kilometres (250 miles) ENE of London. Den Helder is on the coast.

has been well warmed on its long southward journey over a sea surface of at least 7°C (45°F). Our seas are a good defence against winter cold. Likewise they are responsible for the relative coolness of our summers as warm winds from the landmass of Europe are cooled down by the sea, while tropical air from the Gulf of Mexico also undergoes substantial cooling. Even polar winds, during the summer, are less warmed by the sea than they would be if there were a land bridge from Scotland to Iceland.

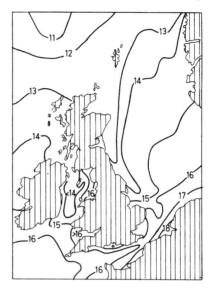

Fig. 4. **Average sea-surface temperatures in August**

The Channel usually has the warmest sea-water in August, about 17°C (63°F); as the more detailed map (Fig. 4) shows,[4] although sheltered bays and estuaries where there is a big rise and fall of the tide, experience much higher sea temperatures, since the rising tide is appreciably warmed by contact with sand and mud. In settled spells it is quite common for the sea surface to reach 20°C (68°F) or more. In these spots it is possible for those taking a dip in the sea to find the water warm enough to make swimming pleasurable, though this is not likely before the middle of June.

The Lancashire coast, the Bristol Channel, the estuaries of Suffolk and Essex, Camber Sands, Chichester Harbour, and Bournemouth Bay are among the favoured localities. These warm patches of water and their mudbanks provide extra moisture to the air as well as extra warmth, sometimes possibly triggering off thunderstorms. Belts of colder water, on the other hand, such as occur off the north-east coast of England, have the opposite effect and inhibit shower formation.

Winds from the Atlantic will generally bring cool, cloudy summer weather to western coasts. The east coast will often fare much better. At Gorleston, in Norfolk, the highest afternoon temperatures often occur with Atlantic winds. As winds from this quarter are commonest, the east and south normally have the best summers on average, but when winds blow from the east over the country, then the west normally has the best weather, with North Cornwall, Dyfed (Pembrokeshire), Lancashire and Cumbria having long hours of sunshine and high temperatures. In August 1947, when easterlies prevailed throughout the entire month, the maximum at Skegness, in Lincolnshire, was 20°C (68°F). Blackpool, well sheltered by the Pennines, averaged 24°C (75°F).

Fig. 5 illustrates the importance of wind direction on weather: easterly winds are bringing fog and low cloud to the east coast from Kent to Northumberland and also

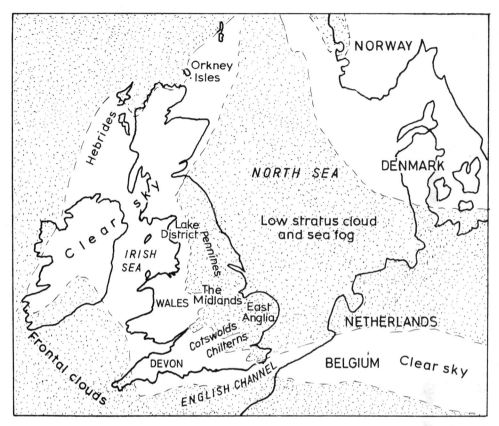

Fig. 5. **14 June 1970. Easterly winds are bringing cloud and sea fog to East Anglia and much of the Midlands, while Wales, Lancashire, Cumbria and much of Scotland are enjoying fine warm weather.** Frontal cloud is approaching across Cornwall, Ireland and the Hebrides.

to much of the Midlands, while Wales and Lancashire as well as most of Scotland are free of cloud.

Sea breezes are local rather than general atmospheric winds (see Chapter 20) and may occur on any day in the warmer half of the year when there might otherwise be only light and variable winds. By the sea-shore, the morning will often start still and sunny; but as the sun climbs higher into the sky the sun's rays warm the land much more rapidly than the sea, so the air over the land becomes warmer than that over the sea. Over the land, the warm air rises and cooler sea air flows in to replace it, so a temperature (and pressure) gradient is set up. By mid-morning, air from over the sea moves inland forming a gentle breeze, which will usually blow at rather less than a right angle to the general direction of the coastline; and, depending on its strength, will tend to keep temperatures in coastal places not far above that of the sea surface. Differences of 6° or 7°C (11°F or 13°F) between resorts in Kent, Essex and London are

Fig. 6. **10 h, 24 April 1977**. Convection clouds over the land have picked out the shape of the British Isles with cumulus clouds. The sea is still cold and winds are light, so there is no convection cloud over the sea. The clouds over the North Sea (*top right*) are associated with a depression over Scandinavia. A warm front from a depression in mid-Atlantic is bringing cloud towards Ireland (*bottom left*).

not uncommon. However, a compensating factor is that sea-breezes often keep places free of cloud, particularly cumulus, which readily builds up inland, especially in the summer, but also at other times of the year as Fig. 6 shows. The actual division between air originating over the land and air originating over the sea will often be

Fig. 7. **13.40 h, 8 August 1962.** Looking south near Winchester. The line of cumulus clouds, with their ragged curtain-like bases, indicates a sea-breeze front, which divides the land air from the sea air. The arrival of the sea-breeze is usually accompanied by a drop in temperature and a rise in humidity and often the clearance of any cloud. Sometimes, sea-breezes are welcome to temper the heat of a summer afternoon.

marked by a line of cumulus clouds (Fig. 7) distinct from the scattered cumulus still further inland. But not all sea-breezes produce fronts. On the landward side the air will often be hazy; on the seaward side visibility will be good. On the north-east coast around Middlesbrough, where air pollution is still considerable, the onset of a sea-breeze from the North Sea will frequently improve the visibility dramatically and bring a tangy freshness to the air, as the smoke is rolled away south-westwards into the Vale of York. From Lincolnshire northwards the sea-breeze will keep afternoon temperatures lower than along the Channel coasts, for the sea is much colder, and maximum temperatures at coastal resorts along the north-east coast can be in the range of 12°C to 14°C (54°F to 57°F) in early summer in sea-breeze situations. By August the sea is warmer and late summer is often pleasantly warm on the Norfolk Broads and at Scarborough and Whitby.

When one feels the need to seek shade, even on breezy days, it is quite likely that the thermometer will have reached 25°C (77°F). Such hot days are relatively rare in Britain because of the cooling influence of the sea around us. On average, Cambridge

Fig. 8.
1 July 1968. Water-lilies speckled with orange blotches of dust transported from the Sahara and brought down to ground by a few large spots of rain falling from the middle atmosphere. The dust fall was widespread from east Devon to Yorkshire, though the Cotswolds were not affected because rain washed away the dust.

has thirteen such days a year; Kew eleven; Hastings four; Armagh and Glasgow two each. At Stornoway in the Outer Hebrides 25°C (77°F) has been recorded twice since records began in 1880. At Lerwick, in our northern isles, the thermometer sometimes fails to reach 18°C (64°F) in a whole year, a value frequently reached by April in the Midlands and not all that infrequently in March. Some say that this is because Lerwick is so far north: it is slightly nearer the North Pole than the southern tip of Greenland, but the northerly latitude is not the main reason for its cool summers. The cool sea and the distance from a large landmass make it almost impossible for warm air masses to penetrate to the Shetlands without losing their surface character. In the same latitude in Canada, 60°N, Fort Smith in the North-West Territories, which is well away from maritime influences, regularly experiences shade temperatures over 30°C (86°F). The record there is 39°C (102°F), higher than anything ever recorded in Britain.

The warmest summer weather occurs in the south-east, nearest to the European landmass, while western and northern parts rarely experience hot days; though what the thermometer says and what one feels are not necessarily the same, for the humidity of the air affects our appreciation of temperature. In general, the highest temperatures will occur with light winds, allowing the air to remain in contact with sun-warmed ground for a longer time, although sometimes the heat is largely imported by the wind. On 1 July 1968 very hot winds from the Sahara brought temperatures as high as 32°C (90°F). The origin of this airstream was shown to many people who awoke to find blotches of orange desert dust covering cars, pavements, roofs, plants and washing. Fig. 8 shows such an example on some water-lilies.

Some of these examples show the effect of the sea on our weather which makes it more equable, damp and liable to frequent change. But we are fortunate not to experience sudden changes as sharp as those affecting the interior of North America. St. Louis, Missouri, has experienced a midnight temperature of 21°C (70°F), followed twelve hours later by −2°C (28°F), or a change from summer warmth to winter cold in hours when one would expect the temperature to be rising. For comparison, one of the largest changes in London was a maximum of 26°C (79°F) on 2 April 1946 to one of 13°C (55°F) the next day, accompanied by a change from brilliant sunshine to steady rain. At the end of January 1956 the thermometer registered 9°C (48°F) at breakfast time, but by noon, with air of Siberian origin, the reading was −2°C (28°F). Both are sharp contrasts but they are barely noteworthy by North American standards. On a daily basis, a range of 28·5°C (51·3°F) has been recorded in August in a valley in southern England[5] and of 29°C (52·2°F) in a Highland valley;[6] this is nothing compared to a fall of 55·5°C (100°F) in twenty-four hours at Browning, Montana,[7] from 6·7°C (44°F) to −48·8°C (−56°F) on 24 January 1916.

The sea does more than just insulate these islands from extreme temperature conditions. It greatly modifies the flow of air blowing over it (as Chapter 2 describes). It must not be thought, however, that the temperature of the sea surface is always like that shown in Figs. 1 and 2. Very large fluctuations of sea-surface temperature occur, particularly off the eastern United States and Newfoundland, with differences of as much as 10°C (18°F) between one January and another, and of 9°C (16°F) between one August and another. These areas are of considerable interest to British weather forecasters because of the great interaction which takes place between a warm sea and cold air blowing over it, and the converse.

Research in the Meteorological Office, using all available ships' observations, has enabled maps to be produced that show anomalies (or differences from the average) of ocean temperature in each five-degree square in the North Atlantic.[8] In the majority of months since 1880, it has been possible to classify the temperature distribution of the sea surface into eight types: three of them warm, corresponding to warmer than average water in the middle, west and east of the Atlantic; three cold, with a similar classification; one type where the ocean is warmer than usual in the north and cooler in the south and another the opposite to this.

Sea-surface temperatures seem most important to the forecaster in winter when the weather is usually windier and the differences between air and ocean temperature are greater, though in all seasons the effect can be considerable. One of the coldest Junes of the century (in 1972, see page 156) was associated with west winds coming off a colder than usual North Atlantic. The mid-Atlantic was warmer, which provided a breeding ground for a series of depressions bringing rain and dull weather with unseasonably low temperatures to much of England. The sea-surface temperature was also partly responsible for the copious rains in the autumn of 1976, when the waters of the English Channel were substantially warmer than usual after a hot and sunny summer, so that evaporation was higher and there was more water vapour available to condense into clouds and rain. September 1995 was very wet in some parts, partly due to the above average warmth of the sea after the hot summer.

If the sea temperature influences weather, the reverse must also be the case. In 1947 and 1963, cold easterlies reduced the surface of the North Sea to just above freezing point, and that in turn reduced the warming influence of the sea on cold winds. Severe gales will also lower the sea-surface temperature by churning up the warmer surface layers and raising colder water from below.

NOTES TO CHAPTER 1

1. Nigel Calder, *The Weather Machine*, BBC, 1974, p. 64.
2. Met.O. 483, HMSO, 1959. Weathership 'J' data, 1967–71, from Marine Climatological Section, M.O., Bracknell. Weathership 'J' is no longer operating.
3. D.W.R. February 1956.
4. Charlottenlundslot, Denmark, average sea-surface temperatures in August, 1905–54.
5. Rickmansworth Frost Hollow.
6. At Tummel Bridge, Tayside, on 9 May 1978. Minimum −7°C to maximum 22°C.
7. *Guinness Book of Records 1996*, Guinness Publishing, 1995.
8. R. A. S. Ratcliffe, 'Recent work on sea-surface temperature anomalies related to long-range forecasting', W. 28, pp. 106–17. R. A. S. Ratcliffe, 'North Atlantic temperature classification, 1877–1970', M.M. 100, p.225.

2 | More about the Sea and Air

Every wind has its weather.

Bacon

The tropical Atlantic, the Arctic seas and northern Europe are the chief sources of the air that the wind carries over us. These three sources all contribute certain obvious characteristics of moisture, temperature and feel to their respective air masses, although important modifications occur on the journey to the British Isles.

Tropical air, as it moves north-east from the Caribbean, the Azores or Madeira, flows across cooler seas. Such air is likely to become more stable, with the air at various levels in the lower atmosphere, at least, flowing steadily and evenly, parallel to the ocean. If the surface layers are cooled to the dewpoint (the temperature at which the air becomes saturated and condensation takes place), sea fog will form. If there is enough wind to create turbulence, or mixing of the lower layers of air, low stratus clouds will be more likely (Fig. 9).

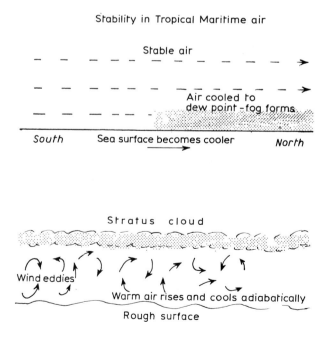

Fig. 9. **Wind blowing over a rough surface causes turbulence.** The rising air cools, and if it reaches condensation level, clouds form. Here stable air results in layer cloud; if the air is unstable, cumulus or stratocumulus clouds may form. Turbulence cloud may form over the sea or land.

Air approaching the British Isles from the polar regions will be warmed by the sea surface, for even in high latitudes the sea is usually much warmer than air from the Arctic, which blows over it, as we have seen. This is true at any time of the year, so that the air will tend to rise vertically, forming large cumulus and cumulonimbus clouds (Fig. 11), since a shallow pool of air which is much warmer than its environment rises. In unstable weather conditions the temperature fluctuates as the rising warm currents are replaced by cold down-draughts.

The prevailing winds over Britain come from the south-west. Sometimes a south-west wind will feel mild and damp in mid-winter, with afternoon temperatures around 10°C (50°F) from Cornwall right across to Aberdeen. This is quite warm enough for a coat to seem unnecessary, if not undesirable, unless it is very windy. On the other hand, with the wind in the same direction the thermometer may read 5°C (41°F). During the day there may be blustery showers followed by a touch of frost at night and in the early morning. Things will feel very different, too, because the lower humidity and the lower temperatures will give the air a polar feel. Some would say a polar smell as well. The clouds will also be different, with their lumpy white turrets gleaming in the sunshine. Obviously, these distinctive examples of weather are caused by air from different sources approaching from the same direction. Any attempt to classify the weather by wind direction alone would be misleading.

During the winter a north wind is usually associated with low temperatures and snow. Sometimes, however, a northerly wind over England does not bring wintry weather. This is because the air has not originated in the Arctic, but has spent some time over the Atlantic, travelling eastwards before turning south. Just the opposite can happen with the wind in the south. Most of us would associate this with mild winter weather. But such a wind can cause surprise by bringing intense cold, as in 1940, when northern France was in the grip of severe conditions. In reality, this south wind is nothing more than a cold easterly or continental air mass deflected northwards.

The descriptions that follow indicate the weather to be expected when one particular air mass is in control, and when there are no fronts or anticyclones near the observer.[1] The term 'air mass' was first used in meteorological circles in 1928 at the Bergen Meteorological Institute to refer to a largely homogeneous mass of air with well-marked characteristics of temperature and humidity. 'Tropical' implies warm air, 'Polar' implies cold. These names are used to distinguish air masses and their sources (Fig. 12). What happens on the margins of air masses, where depressions and fronts are to be found, or where air is stagnating in a high pressure system, will be described in Chapter 3.

Polar Maritime air is the commonest air mass (Table 2), influencing our weather for just over a third of the year at Kew and just under half at Stornoway. That is to be expected since Stornoway is much nearer the origin of polar air. Polar air, as it sweeps south towards the British Isles, has to cross a wide expanse of sea that is exceptionally warm for the latitude (Figs. 1 and 2). Showers, often with hail, are frequent over the seas to the north and west of Britain. At Ocean Weathership 'J' hail occurs on average on sixty days a year. Hail is also frequent over western coasts and

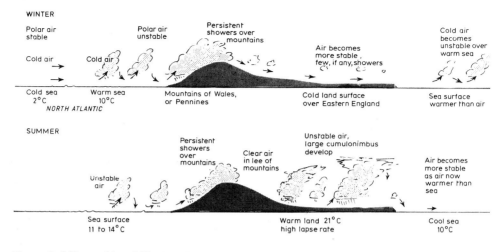

Fig. 10. **Stability and instability in Polar Maritime airstreams.**

Fig. 11. **A distant band of cumulonimbus viewed from the western side of Dartmoor.**
Tavistock lies in the middle distance. The picture was taken about 10 h on 7 November 1976.
It is likely that the shower was triggered by polar air crossing a relatively warm sea. Note
the anvil which is a good indication of a shower in progress.

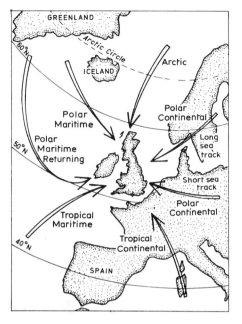

Fig. 12. **Air masses and their sources.**[1]

Table 2 **Percentage Frequency of Different Air Masses**[1]

Air mass	KEW	SCILLY	STORNOWAY
Arctic	6·5	4·2	11·3
Polar Maritime	24·7	27·5	31·5
Polar Maritime Returning	10·0	10·0	16·0
Polar Continental	1·4	0·9	0·7
Tropical Maritime	9·5	13·5	8·7
Tropical Continental	4·7	2·7	1·3
Air in or near anticyclones	24·3	22·1	13·8
Air in the vicinity of fronts	11·3	11·8	11·8

hills, but the hailstones are nearly always very small. Giant hail, the size of tennis balls (Fig. 51), is a lowland and usually a summer phenomenon. Afternoon temperatures will often reach the January average of 7°C (45°F) at Kew Observatory, but night frost will occur if the air becomes still, as it frequently does, in sheltered parts such as the Thames Valley, the Vale of York and the Central Lowlands of Scotland. The colder land will then act as a stabilising influence (Fig. 10) and cool the air from below, so making it less liable to rise and form shower clouds. This is borne out by events, as showers are much less common in the south and south-east in winter than further west, and are quite uncommon at night, except in very cyclonic situations.

At night, during the summer half of the year, showers will often fall in the west and north, with much of the south of England and the sheltered lowlands to the east of the Pennines and Scottish Highlands remaining dry and clear after dark. By day, things are totally different (Fig. 10) since strong sunshine will warm up the ground, heating the air and causing convection currents to rise rapidly into the colder layers of the atmosphere, so that the east will have showers as well as the west. At a few thousand metres, polar air will be much colder than in a tropical airstream, even though the surface temperature may not be vastly different. Put another way, this means that the temperature will often fall more rapidly with height in polar air than in tropical air. Once air has begun to rise, it will continue to do so until it is the same temperature as its surroundings (Fig. 11). Vertical, or convective, currents will begin to rise fairly early in the morning and, frequently before people have left for work, there will be some evidence of this to be seen in the sky. By noon, the sky will be largely cloudy and a few showers may well have begun. This is the sort of weather which must have given rise to the saying: 'It is too bright to last.' The glorious early morning light gives way to showers or rain by afternoon, or at best to a rather threatening sky. Polar Maritime air is likely to be good for landscape photography. It is turbulent and comes from a clean source, and so produces very clear conditions, except in showers.

A variety of Polar Maritime air is called *Polar Maritime Returning*. It is fairly frequent and, as its name implies, is an air mass of polar origin which is returning polewards. First, the polar air sweeps south in the normal way in mid-Atlantic so that its surface layers undergo substantial warming and become unstable; but then the air moves north-east towards the British Isles across a sea which is becoming cooler. The air is now cooled from below by the sea; hence it becomes more stable with layer clouds becoming commoner than cumulus. In winter, the weather it produces will be not unlike that for Tropical Maritime air, with low cloud and drizzle affecting the higher hills, such as the Brecon Beacons and Dartmoor. In summer the cloud will break up away from the west coast, and cumulus clouds may form. However, the air will betray its polar origins as it is still cold and unstable, and some clouds may grow

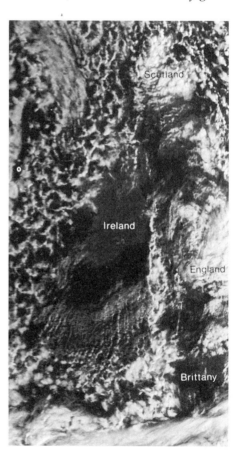

Fig. 13. **A northerly airstream over Ireland.** As the sea is warmer than the air, convective clouds have formed over the seas to the north and west. Once over the land, the air becomes stable after a cool night and remains clear of cloud. For some 80 kilometres (50 miles) off the south coast the sky remains clear of cloud until convective stirring begins again, when the warmer sea causes turbulence. This satellite image illustrates a real example of the principle shown in Fig. 10.

big enough to produce showers and thunderstorms. A typical sequence of weather for this air mass is as follows: the morning starts with a covering of stratus, this gives way to bright periods by mid-morning suggestive of a fine afternoon, but then shower activity may well build up.

Arctic air is usually colder than Polar Maritime air, for it has a shorter sea crossing. In the Shetland Isles and the north of Scotland it can bring severe weather as late as April, since the sea is at its coldest about then around the northern isles and so exerts least warming. Like Polar Maritime air it is unstable and showery over the sea and on windward coasts. It is not the coldest air mass in the south of England (Table 3). In all parts of the British Isles it can bring cold weather with snow to exposed hills and coasts, and the strong northerly winds often associated with this air mass make it feel bitterly cold, in spite of abundant sunshine for the time of the year. Arctic air also has the habit of bringing sudden snowfalls from small depressions, called polar lows, which develop unexpectedly. It was Arctic air which was responsible for the dry powdery snow that fell over the south-east on Good Friday, 28 March 1975, and during the next ten days or so.

Tropical Maritime air comes from the seas around the Gulf of Mexico, the Caribbean, the Azores or Madeira. During the winter the weather will be mild and damp, with much low stratus cloud (Fig. 9). Temperatures during the day will exceed 10°C (50°F) widely, even though there may be little or no sunshine. If the cloud clears and the tropical air stagnates for any length of time, fog may form rapidly during the night. Generally the night minimum, which frequently occurs about sunrise, is not far below the day's maximum. Early morning readings of 7°C (45°F) are not unusual, even if the actual feel of the air is not particularly pleasant as it is very muggy. The presence of this air mass does help to keep heating bills down because of the mild conditions it brings. Also there will be no interruption of transport services due to wintry conditions, although there may be problems because of fog. The arrival of Tropical Maritime air after a cold spell is often very apparent because of copious condensation on walls, glass and other cold surfaces. Any deficiencies in the electrical system of a motor car will be shown up, too, making starting difficult if not impossible. In November 1947, alternating spells of cold northerlies and mild south-westerlies bringing tropical air gave several days of copious condensation. October 1995 saw some very humid air with the outside of house windows becoming steamed up, which indicated that the interior of a house was cooler than the air outside.

Rainfall in Tropical Maritime air will not generally be large in the eastern half of the country; but it is very likely that all high ground in the west will be cloud-covered down to 150 metres (500 feet). London and the Vale of York, for example, may just have intermittent drizzle, while continuous downpours will be likely over the higher hills and mountains of Wales, and lighter rain may fall on lower summits such as Leith Hill, Surrey, and Ivinghoe Beacon in the Chilterns. November 1994 had persistent winds off the mid-Atlantic with above average rainfall in the western half of Britain and less than 50% of the expected rainfall in East Anglia.

Table 3 **Average Maximum and Minimum Temperatures in Different Air Masses at Kew (°C)**

	January		July	
Air Mass	*Max*	*Min*	*Max*	*Min*
Arctic	2	−2	18	11
Polar Maritime	7	2	21	13
Polar Maritime Returning	10	7	22	15
Polar Continental	−1	−4	-	-
Tropical Maritime	11	8	22	16
Tropical Continental	9	5	28	16
All air masses	7	2	21	13

In the summer half of the year, Tropical Maritime air will be close and humid on western coasts. Snowdonia and the more westerly Scottish Highlands will frequently have rain because of orographic uplift (see Chapter 4). If the air is very stable, extensive banks of fog may form (as in June 1976) along the Irish Sea and Channel coasts.[2] Inland, there will probably be bright intervals and quite long spells of sunshine along the east coast. If the airstream is more turbulent, low cloud rather than fog will be widespread, penetrating some distance inland, especially at night, and often reaching right across the country to parts of the east coast by morning. The sun will usually burn off the low cloud in places in the lee of high ground so that fine bright weather with a hazy blue sky may occur. Patches of cirrus and fragments of low stratus will provide a changing cloudscape. The best of the weather may well be found in the Vale of Evesham and the Midlands where afternoon temperatures will reach 21°C (70°F) or more. Other favoured spots include much of the east coast from Kent to Scotland and around Torbay in the shelter of Dartmoor. Because the air is humid and oppressive many people will think that a thunderstorm must be imminent. But storms are comparatively rare because of the stability of the air, and so large clouds do not form readily except near fronts. Sometimes Tropical Maritime air produces warm weather with abundant sunshine, especially in the south-east.

For most of the time the air which influences our weather has Atlantic origins. The two remaining air masses to be described originate over the land and have very different qualities. They both bring dry conditions, and it would be reasonably accurate to say that they are responsible for most of our more memorable spells of weather, except for severe gales.

Polar Continental air has its origins over European Russia and Siberia, and its onset during the winter usually means a severe cold spell for England and Wales, though not necessarily for Scotland. Indeed, in January 1963 mean temperatures were several degrees lower in the south of England than in north Scotland, as Fig. 81 shows. The reader will recognise that the table of temperatures for Heathrow and De Bilt illustrates the extreme conditions which can occur in Polar Continental air (Table 1). Fig. 14 shows an example of Polar Continental air in 1972 during a short but intensely cold four-day cold spell. In January 1987 a similar pressure pattern, with the high

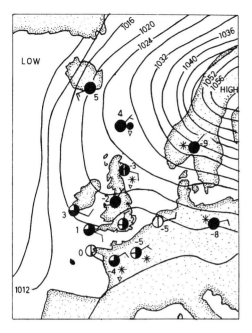

Fig. 14.
Noon, 30 January 1972. Polar Continental air between 28 January and 1 February. Temperatures fell to −17°C (1°F) at Edinburgh and to −18°C (0°F) at Warsop, Nottinghamshire, and East Malling, Kent. By the evening of 1 February the short cold spell was at an end. A similar pressure pattern gave severe cold to most of the British Isles in January 1987.

extending a little further west, produced some of the coldest days of the century, and maximum temperatures as low as −7°C (19°F) as far west as Devon.

The very cold air leaving the other side of the North Sea will often be dry and stable, and the weather fine. Before long the warming influence of the sea, which is at a temperature of 5°C to 6°C (41°F to 43°F) in an average winter, will cause turbulence cloud to form in the lowest layers, sometimes producing a complete covering of stratocumulus. Inland, the cloud cover will be more broken, while western coasts will often have 'perfect' weather with some night frost. Any snowfall will usually, though not always, be light and barely enough to cover the ground (see page 197), except on some of the highest hills such as the North Downs behind Dover and Canterbury, and the eastern slopes of the Pennines, and the Yorkshire and Lincolnshire Wolds. The air will have acquired little moisture from the sea and its origin and temperature will ensure it cannot hold much.

Sometimes Polar Continental air will approach from the south-east taking the shortest sea crossing. When this happens, the cold will be severe at night and even during the day, as occurred in February 1956 and 1986. Except in some hilly parts snow was not a problem for transport.

On yet another track of this air mass, cold air sweeps across Scandinavia and turns south-west towards Britain. The sea journey, though nowhere so long as that made by Polar Maritime air, is sufficient to cause instability, so that large shower clouds build up, giving frequent snow showers in the eastern half of the country, including London.

Between the middle of May and September, or even October, Polar Continental air (now better perhaps called Temperate Continental air because it originates over a warm land mass outside the tropics) will be stable and dry as it leaves the mainland. In summer, the sea is now cooler than the air; and so the lower layers in contact with the North Sea will often be cooled enough for fog or low cloud to form (Fig. 169). This low stratus cloud is frequently referred to as 'haar' in eastern Scotland, where it can sometimes be very persistent, reducing visibility in Edinburgh to below 50 metres (160 feet), even in August. Around London and in the Midlands the dull low canopy of cloud will usually break up by mid-morning unless the wind is strong. There will be little or no cloud in Lancashire, the Cardigan Bay shore, the Lake District and much of west Scotland.

Tropical Continental air is the least common of all the air masses that affect our weather. It most frequently approaches from the south or south-east in summer. On its journey from the Sahara or the Near East it crosses the Mediterranean which is a warm sea. But such moisture as the air possesses will in all probability be deposited over the Alps, the air eventually reaching Britain as a warm dry stable airstream in its lower levels. Unlike Tropical Maritime air, it will not cause much fog in the English Channel, since the cool sea hardly has time to cool the surface layers of the air to their dewpoint (see page 12), as the sea crossing is so short. On the other hand, if the warmer air penetrates farther north, fog will be more probable over the Irish Sea or along parts of the west coast, where the sea-surface temperature is apprecia-bly lower. Fig. 102 shows temperature, wind direction and weather in a Tropical Continental air mass. Many people tend to think of a British summer as 'three fine days and a thunderstorm'. Within the dry tropical air, storms are unlikely; but they can be prolonged and violent where the hot air comes into contact with cooler air such as Tropical Maritime: the cooler air, being denser than the very hot air, will undercut it and lift it, and the cooling of the hot air will then soon lead to the forma-tion of thick clouds and heavy rain.

From the information shown in Table 2 we see that for about two-thirds of the year our weather can be classified according to air masses. Months or seasons which have departed greatly from the average will have had an excess of one particular air mass, such as February 1956 and 1986, January and February 1963, October 1974, January and August 1975, June, July and August 1976, June 1978, July 1983 and November 1994. All of them had winds from the Arctic or from the Continent; only January 1975 and November 1994 had a predominance of Tropical Maritime air. So it is usually continental air which brings memorable weather, except for the worst gales which tend to be associated with Atlantic air.

It should be possible for the reader to identify which particular air mass holds sway on most occasions. For the remaining third of the year, the weather does not fit into the categories described, as things are very different when air masses come into contact with one another to form depressions and fronts.

NOTES TO CHAPTER 2

1. A.A. Miller and M. Parry, *Everyday Meteorology*, 2nd edition, Hutchinson, 1975, pp. 98 and 113. J.E. Belasco, 'Characteristics of air masses over the British Isles', *Geophysical Memoirs*, No. 87, M.O., Bracknell.
2. June 1976 was a good example of contrasts. The north-west of Scotland had a very dull month with winds off the Atlantic.

3 | Highs and Lows

From a satellite photograph (Fig. 16), a low-pressure system can usually be identified as a circular whorl or vortex-shaped pattern of cloud, with a long straggling tail representing a front (the boundary between two air masses) extending hundreds of kilometres from the centre, where polar air meets tropical air. It is the interaction of tropical and polar air, the Coriolis Force and Jet Stream or strong high-level winds, which helps to start off this large-scale vortex. The ideal area for the birth of a depression appears to be the North Atlantic, particularly between Newfoundland and Iceland, though they can form anywhere: over the Bay of Biscay, for instance, or the North Sea.

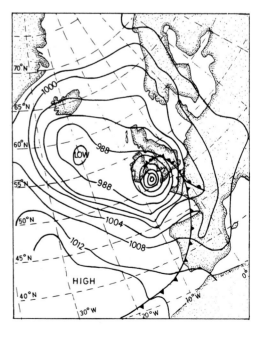

Fig. 15. **Noon, 2 September 1974**

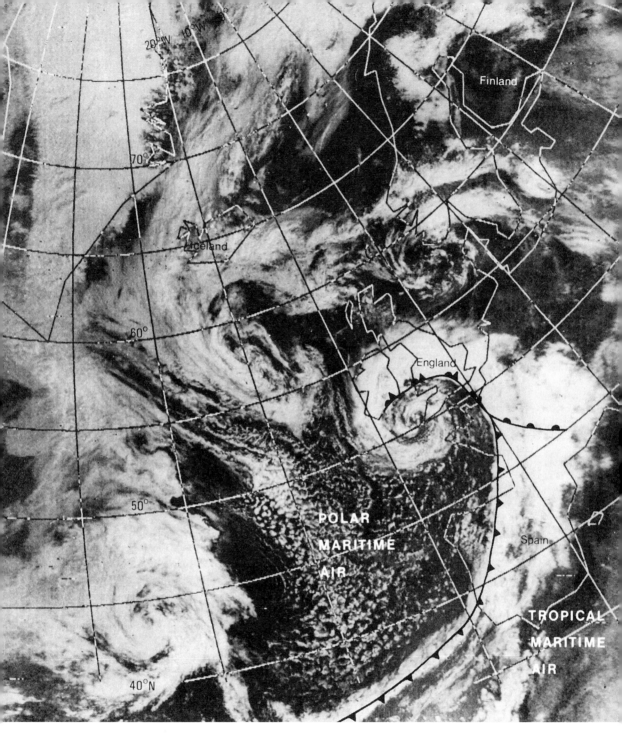

Fig. 16. **10.39 h, 2 September 1974.** A deep depression nears the approaches to the Bristol Channel. There are severe gales on the southern side of the low. Fig. 15 shows the synoptic situation.

The entire cyclonic system cannot be seen by a person on the ground, as it is far too large, but sometimes the banding of clouds in the sky suggests that they could be part of a big vortex. Even before there were high-flying aeroplanes and satellites to take pictures, Bjerknes and others[1] at the Bergen Meteorological Institute, in the 1920s, correctly deduced the shape and form of the depression, especially with regard to fronts and air-mass boundaries. Areas of high and low pressure can also be seen as entities, a pattern of lines on a weather map, with the low having a fairly tight concentric pattern compared to the high. But before considering the nature of high and low pressure systems, it is necessary for us to investigate the vital part played by the barometer in studying weather.

Weather maps as we know them today could not be compiled until the invention of the electric telegraph made it possible to obtain such information as the barometer reading and the temperature and state of the weather from widely scattered places, and to assemble this data quickly and plot the information on a map.

In 1849 *The Daily News* initiated an experiment in weather telegraphy. The sequel was the type of map shown in Fig. 17, which was on sale at the Great Exhibition in 1851 at one penny a copy. The idea of drawing lines of equal air pressure, or isobars, on a map had not yet dawned on us, and it was not until the autumn of 1863 that the earliest weather charts of western Europe appeared with isobars. These maps were produced and organised by Le Verrier in Paris. He was, according to Sir Napier Shaw, the originator of international weather telegraphy.[2]

Today, fax, telephone, teleprinter, radio and radar (for the detection of storms) are vital links in the transmission of the data to the forecasting office at Bracknell. Information also comes from satellite pictures which give valuable help in locating and monitoring the development of depressions or bad weather systems out in the Atlantic.

Weather stations sending information to the forecasting office for the preparation of charts and forecasts are known as synoptic weather stations. They are usually found at ports, coastguard stations, airports and military bases, and anywhere else people are available around the clock to take the readings and forward the information. Data sent include: present and past weather, rainfall, cloud type and amount, pressure and pressure changes, temperature, humidity, wind speed and direction, visibility and the state of the ground, whether snow-covered, frozen, flooded, damp or dry. The pressure readings are plotted on a specially prepared base map, along with as much of the other information as the scale of the map will permit. Lines can then be drawn linking together all the places on the map with the same barometer reading. The resulting pattern of isobars will show high and low pressure centres, troughs and ridges, which can easily be recognised. From the pattern of isobars certain elements of the weather can be forecast, particularly wind direction and strength. Where the isobars are close together on the map, the pressure gradient is said to be steep and the winds will be strong (Fig. 18). An analogy is the flow of water down a slope and of air down a pressure gradient. Where the ground is steeply sloping, contour lines on a map will be close together and the flow of water down the hill

Fig. 17. **Weather map from the Great Exhibition of 1851**

will be rapid. But beware! The analogy is not a good one until the effects of the Coriolis Force are allowed for. This is the force created by the rotation of the earth. It greatly affects the flow of air but is fairly insignificant in its effects on the movement of water over a small area. As a result of the Coriolis Force, air does not flow from high pressure to low directly across the isobars, but at an angle. As mentioned, wind speed and direction can be estimated from the pressure map, especially over the open sea where the estimation is likely to be quite accurate. Over the land, however, allowance must be made for (a) the sheltering effects of hills, (b) the drag on the wind by trees and buildings which tends to reduce the theoretical speed considerably (this factor can be allowed for in the forecast), and (c) gustiness, which is more of a problem (Chapter 22).

Regions of convergence where air currents meet can also be identified with the help of pressure maps.[3] The location of these zones is of great interest to the forecaster since they are the areas where rain or snow is likely.

So the barometer is essential to the construction of the weather map and the preparation of the forecast. The better kinds of mercury, or aneroid, instrument in the possession of many a family could play a part in the weather forecast, if the readings could be transmitted to the forecaster in time, but the majority of us would prefer not to have the bother and responsibility of sending in the data each day, let alone at three-hourly intervals during the night as well as by day. Nor, for that matter, would the forecaster achieve an even coverage of data. There would be too much from the populated parts and too little from sparsely peopled places. To safeguard the quality and continuity of the information, stringent standards are applied at synoptic weather stations.

The barometer hanging on the wall can measure the air pressure perfectly accurately. It cannot show the distribution of air pressure. Its value as a forecasting instrument in isolation is limited until its shortcomings are appreciated. The Victorians and Edwardians had a vogue for ornately carved mercury barometers with a narrow glass tube, inverted in a small reservoir of mercury. Some of these instruments have considerable value today as antiques. When air pressure rises, some of the mercury is forced up the tube making the level rise, hence the saying, 'the barometer is rising'. The aneroid is a smaller and more portable instrument containing a box, or metal coil, from which most of the air has been excluded. As pressure falls, the box or coil expands through pressure changes, and by attaching a suitable arrangement of magnifying levers and pointers we can read off the air pressure from the dial. The mechanism is usually hidden behind the dial, but round the face are calibrations, in inches and millibars or centimetres, plus inscriptions such as 'Set Fair', 'Fair', 'Changeable', 'Rain', 'Much Rain' and 'Stormy'. While it is generally true that the higher the pressure the better the weather, exceptions abound. A reading of 30 inches or 1016 millibars may well be associated with fine cloudless weather on one occasion and with overcast skies, rain and cold winds on another.

On 16 November 1976, for example, many people phoned the London Weather Centre to inquire whether their barometers were misleading them as high readings of 1025 mb (30·3 in) were general. These are mainly associated with fine weather, yet rain was falling in many parts of the country. It came from a trough of low pressure between two anticyclones. The opposite situation occurred in the south of England about a fortnight later, when clear sunny weather prevailed with a very low barometer. A possible explanation is afforded by the satellite photograph (Fig. 16), which shows bands of thick cloud extending from the centre but with a cloud-free zone actually in the middle of the low. Another example of a high barometer reading (1032 mb, or 30·47 in) being accompanied by rain occurred on 25 January 1992, when between 10 and 15 mm (0·4 and 0·6 in) of rain fell over much of south-east England as a cold front penetrated the high.

Over the British Isles a few hours of bright clear weather quite often occur just to the rear of a depression, due to subsidence behind the general uplift at the frontal system, though such fine conditions tend to be local and liable to a sudden ending.

In the previous chapter descriptions of each air mass were given. Some air masses, notably Polar Maritime and Tropical Maritime, are frequent, while others such as Polar Continental are uncommon and may not directly influence the weather significantly in some winters. Which air mass reaches this country depends on the distribution of air pressure. This in its turn is a result of the various complicated physical forces acting on the earth's atmosphere, the ocean, the land (in particular the Plateau of Tibet, Greenland and the Rocky Mountain system) and polar ice. In other words, we can explain why Polar Continental air was virtually absent between 1972 and 1975, and in several more recent winters, notably 1989 and 1990. It was because high pressure never developed over Scandinavia or to the north of Scotland. Six successive Januarys, 1988–93, were without snow in east Surrey. The lack of snow has a connection with the displacement east of the Icelandic low. Behind it (to the west) a secondary developed, thus preventing polar winds from reaching southern England. But we cannot explain why this happened, nor, for example, why high pressure has developed in some years and not in others. Who knows? It may be as well if the human race never discovers the answer (see Chapter 24). What happens in the upper air has considerable effect on the surface weather. The temperate-latitude jet stream, a strong high-level wind, moves in a series of loops around the northern hemisphere and low pressure tends to develop on the eastern side, where the jet turns polewards.

Research has shown that lower than normal air-pressure tends to occur east of colder than normal sea-water, while higher pressure tends to occur east of warmer than normal sea-water. The record warmth of 1975 and 1976 was the result of persistent south or south-easterly airstreams maintained by an extensive cyclonic system in the eastern Atlantic, just east of a pool of cold water, and high pressure over or near the British Isles. Deviations from the average in sea-surface temperature (S.S.T.) often cause depressions and anticyclones to form in different places from usual.[4] But it also seems as if the position of the jet stream is influenced by the S.S.T.

The diagrams (Fig. 18) show the path of air in and around a depression and an anticyclone in plan and in section. In the low, the air spirals inwards towards the centre (an obvious feature on many satellite pictures of clouds) so that it must even-

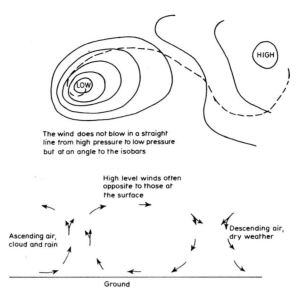

The wind does not blow in a straight
line from high pressure to low pressure
but at an angle to the isobars

High level winds often
opposite to those at
the surface

Ascending air,
cloud and rain

Descending air,
dry weather

Ground

Fig.18. **The flow of air in and around a depression and an anticyclone**

tually rise. Thus the warmer, less dense, air rises over the colder, heavier air. This cools the warm air and leads to precipitation (rain or snow). The air currents do not normally rise vertically, but at a gradient of 1:150 to 1:400 (Fig. 19). Clouds will therefore be of the layer type. A layer of cirrus up to 400 kilometres (250 miles) in advance of the warm front at the surface is followed by altostratus and nimbostratus, which are grey, shapeless rain clouds. The cold front and occlusion front are explained in Figs. 20 and 21.

In the anticyclone, the air is usually converging at high levels. It eventually sinks back to the ground and spreads out, in the northern hemisphere, in a clockwise direction. It thus becomes warmer and drier adiabatically, that is, the air becomes warmer by descent and compression. A bicycle pump illustrates the principle of adiabatic warming. Just try touching the connection after a vigorous effort has been made to inflate a tyre: it will be very hot at the point of maximum compression. Places influenced by high pressure are often free of cloud because the air is dry, due to the descending air mass, which stifles some or all of the convection currents rising from the ground. That explains why some days remain cloudless, in spite of plenty of evidence that there are convection currents present, e.g. heat shimmer and mirages seen over hot surfaces such as roads, mud flats and cornfields. In winter, intense radiative cooling will frequently lead to the formation of fog and frost (page 133). Another winter phenomenon associated with anticyclones is a gloomy day with a layer of stratus or stratocumulus covering the entire sky, often with a considerable accumulation of pollution-products in industrial areas. Here, turbulence near the ground is just sufficient to lift the air high enough for it to condense and form cloud, below an inversion of temperature which effectively prevents any further convection.

The weather associated with a depression is shown in Fig. 22. It is common for a low to travel on a north-easterly course off the Hebrides, shown by track C in Fig. 23. When this happens, southerly and south-westerly winds on the south-eastern side of the depression will bring Tropical Maritime air to all parts of the country. Places in the south may well escape the bad weather, being a long way from the centre, or there may be several hours of continuous rain as a trailing cold front moves across.

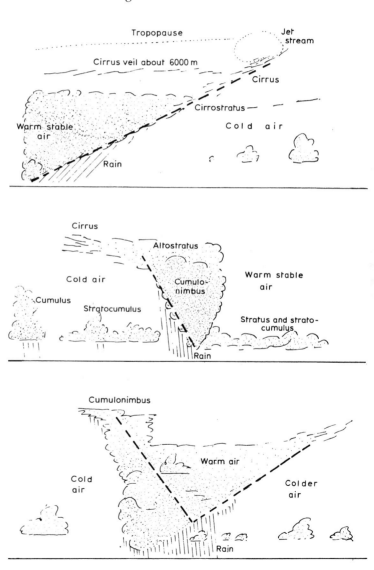

Fig. 19. **Cross-section through a typical warm front.** Bands of cirrus in the western sky are often the first signs of the approach of a warm front. Gradually the cloud thickens into a layer of cirrostratus, altostratus and nimbostratus. From the low cloud light rain or drizzle will fall. In winter, there is a steady rise of temperature and any snow will turn to rain. During the summer, the rise of temperature with the passage of the warm front is less obvious on some occasions, as the sunshine in advance of the front will often make it feel warm, though it may be quite cool in the shade. Occasionally warm fronts produce a lot of rain in the lowlands, if they move slowly or if the warm air is unstable. Sometimes the temperature contrast at the surface across a front is not large; but it may be substantial at a few thousand metres.

Fig. 20. (*middle*) **Section through a typical cold front.** After the passage of the warm front the sky will frequently remain cloudy in the warm sector, with drizzle and orographic rain over the mountains in the west. Signs of the approaching cold front are thickening cloud and a period of moderate rain. The rain often ceases quite abruptly, much of the low cloud clearing away, though showers tend to build up behind the front if the air is unstable. In all seasons, the cold front brings air which is much fresher and is often a welcome relief from the damp air it replaces. If the tropical air, which is undercut by the colder air, is unstable, then heavy rain is possible, as the warm moist air is lifted to great heights.

Fig. 21. (*bottom*) **Section through a typical occlusion front.** Often the colder air behind the depression catches up with the cold air in front. The warm air is then lifted off the ground, or occluded. Occlusions sometimes give a lot of rain.

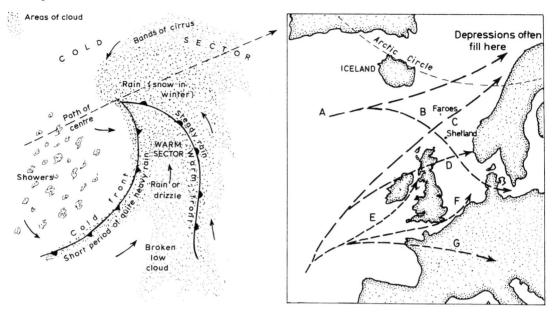

Fig. 22. (*left*) **Typical weather and cloud patterns associated with a depression**

Fig. 23. (*right*) **Some typical depression tracks**

In the Midlands a sequence of winter weather over twenty-four hours, as a deep depression moves north-east towards the Faeroes, may be as follows: a slight frost before midnight in the polar air, in advance of the warm front, is followed by a rapid increase of high cloud, a rise of temperature, thickening cloud and rain as the warm front approaches. Behind the front there will be low cloud and drizzle with perhaps a short period of heavier rain before cooler, showery weather arrives, with a slight frost likely by early evening. A few flakes of snow, or some sleet, may be seen by an observant weather-watcher before the warm front arrives, and after the cold front.

Several times a year a depression will travel on a totally different course, approaching from the south-west and subsequently moving eastwards along the English Channel or into northern France (tracks F and G). When this happens, Tropical Maritime air will not pass over Britain at all. In summer this may mean several hours of continuous rain over southern England with afternoon temperatures below even 15°C (59°F). From the Wash and mid-Wales northwards, pressure will be higher and the weather better, with the best conditions often found along the Lancashire and North Wales coasts, sheltered from east winds; in fact, a wet day in Bournemouth with a north-easter probably means a fine, or reasonable, day in Blackpool. On 4 August 1974, a small depression moved along the Channel from Brittany, bringing heavy rain and thunder and a maximum of only 12°C (54°F) at several places in the south, a value hardly expected in a peak holiday month but not so rare in January. On the same day the Lancashire coast had no rain, eight hours of sunshine and a maximum of 20°C (68°F), not hot but respectable. A low taking a similar track between November and mid-April is likely to produce rain or sleet; or snow if the air is cold enough, and a covering of snow over the higher moors.

Low pressure over the southern North Sea or Scandinavia and high pressure on the Atlantic will produce northerly winds over much of the British Isles. Places sheltered from the north by mountains, such as Glasgow, north Lancashire and South Wales, will do quite well for sunshine and often be the warmest places, while the eastern half of the country, including London, will be cold and miserable. Such a day was 11 June 1995, one of many such days in a long run of north winds. No sun in London; but five hours in Glasgow and seven in Tiree, and not so much wind.

Another very common depression-track lies just to the south of Iceland (track A). Tropical Maritime air will usually cover all of the British Isles, bringing much rain to the north-west, though sometimes southern England may have anticyclonic weather, with night frost and fog in winter, or a heatwave in summer. Sometimes lows will fill or stagnate near Iceland or the Norwegian Sea as in June 1976. While the south was enjoying great heat and sunshine and suffering a record drought, the Hebrides had their dullest June on record.

Notable gales and storm surges have occurred when depressions have followed track B. The north-west winds in the rear of the depression pile up the water in the southern part of the North Sea. That happened in 1928, 1953, 1976 and 1978 (Chapter 19). Sometimes northern Scotland can be very wintry when a low crosses the middle of the country, drawing in very cold continental air from Scandinavia or the Arctic as in 1955 and 1978. A low-pressure centre in the Irish Sea will bring heavy downpours to the north-west of England and, if the depression is deep enough, severe gales as well. The tracks shown are extremely generalised and depressions sometimes follow irregular and zigzag paths, which can make a mockery of the forecast.

Anticyclones are mostly regarded as desirable things to see on the weather map or hear about on the weather forecast. And so they are for those going on holiday, providing the high is in the right place. Which brings us to the question, 'Which is the right place?' Possibly the nearest to an ideal situation for a summer anticyclone would be over the northern Pennines, so that fine weather occurs nearly everywhere, although Tropical Maritime air may give cloud in the north-west Highlands. Fig. 24 shows an anticyclone over northern Scotland. Westerly winds are giving cool weather in the outer isles, while easterlies are bringing low temperatures to the east coast. Continental air in the south is free of cloud and temperatures there are pleasantly warm, although 15°C (59°F) and cloud at Scilly suggests that the anticyclone is not strong enough to keep a trough of low pressure out of the south-west of England. If the high were to move south to the Midlands, Tropical Maritime air would also move south and then places from east Scotland to Devon would have fine weather, but it would most likely be cloudy from west Wales northwards. Over the higher moors and mountains there could be rain, orographic or frontal, in the Hebrides and Orkneys. Should the opposite happen and the anticyclone move north rather than

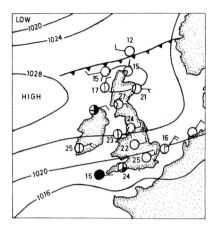

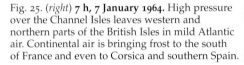

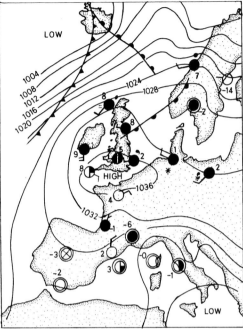

Fig. 24. **18 h, 17 July 1972.** High pressure across Scotland

Fig. 25. (*right*) **7 h, 7 January 1964.** High pressure over the Channel Isles leaves western and northern parts of the British Isles in mild Atlantic air. Continental air is bringing frost to the south of France and even to Corsica and southern Spain.

south, most of Scotland would have blue skies and sunshine with high temperatures in the glens, in all probability. From Suffolk to Devon the weather would be cooler with strong easterlies if the pressure gradient was steep; otherwise the weather would continue settled. A low forming over the Channel, with its threat of rain, may make for dismal weather in southern counties. It is possible to work out what sort of weather will be likely if the anticyclone moves not north or south, but in any other direction. A particularly common location is west of Ireland, permitting a generally north-westerly flow of air. If the wind is strong then the summer weather is rather cool and disappointing.

The best place for a winter anticyclone depends on whether one likes mild 'open' winters without frost, or not. For lovers of mild winters, then, the high should remain over France or farther south, as happened in January 1964, a sharp contrast to the previous winter (Fig. 25). Tropical Maritime air was persistent particularly over western and northern parts of Britain. Generally frost was well below average frequency, but Spain and France were colder than usual and returning British holidaymakers found it became much milder as they travelled north to the Channel ports. A similarly placed high in January 1975 gave a record mild month with daffodils in full bloom by the 15th in London parks. The mild winters of the late 1980s and early 1990s also saw high pressure to the south.

The high, sitting overhead, which gives such good weather nearly everywhere in the summer, is not so welcome in the darker months. During the winter an anticyclone centred over or near Britain will give either a dull quiet spell or (fortunately less commonly) a very frosty, foggy spell. As a generalisation anticyclones, or high-pressure areas, are bringers of good weather; but in our towns and cities severe air pollution may occur. Perhaps the adjective *good* should not be used and *settled* instead. Because winds are usually very light in an anticyclone, pollution products are not very easily dispersed. As well as the lack of wind there is usually a temperature inversion (see Chapter 12), meaning that temperature does not fall with height in the lower layers of the atmosphere. Rising currents of smoke and exhaust gases are trapped below the warmer air at a few hundred metres. Just before Christmas 1994, in an anticyclonic foggy spell, air pollution in London reached levels that were injurious to health. Similar problems have occurred in the summer, with strong sunlight making the situation worse, because of chemical reactions.

If the sky remains clear of cloud during the night, and the high permits moist air of tropical origin to stagnate, persistent fog will form. It is true that as the high moves overhead there will often be a fine sunny day. But by the next morning fog and frost will be widespread, except around favoured coasts. Because of the low angle of the sun there will not be sufficient warmth to cause enough convective stirring to disperse the fog, which may last all day in low-lying and industrial areas, as happened in Glasgow in November 1977 and in London in the Decembers of 1952 and 1962. Outside the foggy areas the weather will often be fine and frosty; and sunshine totals

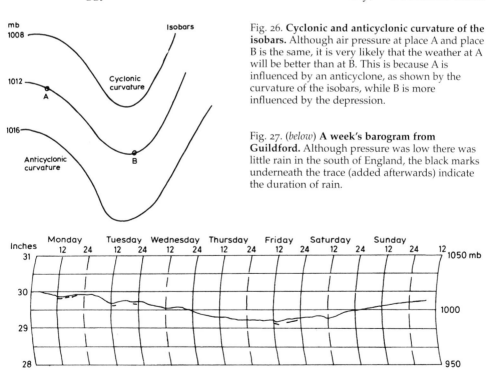

Fig. 26. **Cyclonic and anticyclonic curvature of the isobars.** Although air pressure at place A and place B is the same, it is very likely that the weather at A will be better than at B. This is because A is influenced by an anticyclone, as shown by the curvature of the isobars, while B is more influenced by the depression.

Fig. 27. (*below*) **A week's barogram from Guildford.** Although pressure was low there was little rain in the south of England, the black marks underneath the trace (added afterwards) indicate the duration of rain.

will approach the maximum possible for the time of year. Favoured places include the high ground in the Cotswolds, Hindhead in Surrey, Shaftesbury in Dorset, the Pennines and the mountains of the Lake District, which will be clear and afford an intriguing view over the top of the gleaming white 'cotton wool'.

The descriptions and examples given are generalised, but there are many exceptions, since high- and low-pressure systems are not symmetrical or of a standard shape and the position of troughs and ridges can alter the situation considerably. There is no ideal place for a high to give really good weather everywhere, nor for a low to give the opposite. In both 1947 and 1963, north-west England and western Scotland had magnificent weather for winter, with crisp cloudless days; while the rest of the country was experiencing deep snow and very little sunshine. Then the high pressure was often between Scotland and Iceland or over Scandinavia. Nor, from one depression, are several hours of rain probable in all of the five countries which make up the British Isles.[5] A possible exception would be a track from southern Ireland northwards along the Irish Sea.

So one barometer on its own is not of great use in forecasting the weather, for it is the distribution of pressure, as much as its magnitude, which matters, as the examples show. The curvature of the isobars is also important (Fig. 26), while the barograph trace (Fig. 27) shows how misleading the barometer can be unless used with care and a check on the clouds. On the other hand, to encourage some faith in the barometer, here are some very useful sayings about its behaviour (Fig. 22).

> When the wind backs and the weather glass falls
> Then be on your guard against gales and squalls.[6]

A backing wind changes from south-east to north-east which means that a depression is coming very near and may well pass a little to the south so that the broadest band of bad weather can be expected.

> Should the barometer continue low when the sky becomes clear
> after heavy rain, expect more rain in a very short time.

This means that the depression has not moved very far away so that no lasting improvement will take place until the barometer rises.

> Long foretold, long last;
> Short notice, soon past.

A slowly rising barometer indicates approaching high pressure. As anticyclones move only slowly in comparison with depressions, the value of the first line of this piece of weather lore can be appreciated. A quick fall of pressure means that the low will probably come and go quickly, taking the sequence of fronts with it as it travels. A quick fall and a rapid rise may well mean that the bad weather will come and go quickly and that tomorrow will be fine, or at least not too rainy. The day after tomorrow could easily see a rapidly falling barometer once more, or, as the forecasters say, 'continuing unsettled'.

Extremes of air pressure for the British Isles are 1054·7 mb (31·15 in) at Aberdeen on 31 January 1902, and 925·5 mb (27·33 in) at Crieff, Tayside, on 26 January 1884.[7] A notably high barometric pressure of 1047 mb (30·92 in) occurred over eastern England on 24 December 1962, in association with an anticyclone over Denmark. Nearly two months of severe weather ensued (see page 197). The same pressure was recorded on 30 November 1980, but this time the winds were of Atlantic origin during the days following and the weather of December 1980 was not unusual. During the great gales of 1 February 1953 and 2–3 January 1976 (Chapter 19) pressure fell to about 970 mb (28·7 in) over the North Sea, and it fell to 960 mb (28·35 in) in the storm of October 1987 and to 950 mb (28·05 in) in January 1990. To the north-west of Scotland on 10 January 1993 there was a remarkable record-breaking low-pressure reading of 916 mb (about 27 inches of mercury) and way off the scale of most barometers.[8]

Generally speaking, changes in air pressure at the earth's surface have no effect on our daily well-being unless we are unlucky enough to find ourselves in the middle of a tornado, where air pressure is very low and suction effects operate (see Chapter 9), so that objects may quite literally be levitated. However, the passage of a deep depression can cause problems in mines, where methane is present. A rapid fall in air pressure means that much methane is emitted and at Allerton Colliery underground operations were stopped as a safety precaution on 24 March 1986. Pressure had fallen 14 mb (0·41 in) in twelve hours. Since the connection between methane emissions and air pressure has been established, there have been fewer unexplained mine explosions.

NOTES TO CHAPTER 3

1. Prof. J. Bjerknes, Bergen Meteorological Institute, and his father Prof. V. Bjerknes. They initiated the plotting of fronts on weather charts and the concept of air masses.
2. Sir Napier Shaw, *The Drama of Weather,* 'The weather-map presents the play', Cambridge University Press, 1940, pp. 242–53.
3. On the earliest British weather maps isobars were drawn in inches: 29·9 in, 30·0 in, etc. The internationally agreed unit is the millibar, used by the Meteorological Office since 1914. The millibar is a scientific unit based on the centimetre-gram-second system (cgs). Approximately 29 in of mercury = 982 mb, 30 in = 1016 mb, 31 in = 1050 mb. Also called hectopascal (hPa), as used by the W.M.O.
4. R.A.S. Ratcliffe, W. 28, pp. 106–17.
5. T.M. Thomas, 'Precipitation in the British Isles in relation to depression tracks', W. 15, pp.361–73.
6. Richard Inwards, *Weather Lore,* S.R. Publications, 1972.
7. Ingrid Holford, *Guinness Book of Weather Facts and Feats,* Guinness Superlatives, 1982.
8. E. McCallum and N.S. Grahame, W. 48, pp. 103–7.

4 | The Elements of the Weather

Let us now consider the various elements, such as rainfall, hail, thunder, heat and cold, snow, sleet, fog, wind and sunshine, that make up the weather. We shall see how and when they occur, and how they vary in different parts of the country.

Mountains, which stretch above the tree line to almost permanent snow; bare rain-swept *moorlands*; narrow *valleys*; sheltered *plains* and *estuaries*; high rocky *coasts* with stunted trees and small *islands* harbouring sub-tropical plants: they all have their own weather. The same depressions move over them all, the same anticyclones rest benignly on them; but each takes something to itself. A depression passing over Snowdonia gives very different weather there from in flat Lincolnshire.

What is the main feature of British weather? Many visitors to Britain are impressed by the greenness of the countryside. Nor does the holidaymaker returning from the Costa Brava or Greece fail to notice the green lushness of the land. Britain is a damp rainy country by comparison. Rain usually fills our rivers, gives us our water supply, and vitally affects the growth of our crops. It can make a green glory of the country-side in spring, or a sodden mess of a cricket match; it can seriously upset the finances of clubs and societies trying to raise money at an outdoor function. Is Britain really a rainy land or do we extend the national habit of denigrating our country to its weather? Ignoring downpours over mountains we find that London's average annual rainfall is 599 mm (23·6 in) at Kew Observatory, less than Rome with 915 mm (36 in), Venice with 770 mm (30·3 in) and Nice with 862 mm (33·9 in).[1] Many of us think of Rome and Nice as having less rain than London. Many more seem to think it is always raining in Manchester. It is time to examine the facts.

Weather Records

But why keep weather records? Napoleon's troops were defeated by General Winter in 1812, by the intense cold near Moscow. There were no records available to point out the dangers beforehand. The British were also defeated by the cold in the Crimean War. Travellers' tales were not good enough. Weather records were inade-quate in East Africa in 1946, when many hectares were planted with groundnuts to help world fat supplies, at a cost in 1990s prices of hundreds of millions of pounds.

The crops failed because the rains failed: the records were not lengthy enough to show huge variations from the average, as part of the normal pattern.

Keeping accurate records of rainfall and temperature enables comparisons to be made between places. This gives an indication as to the possible crops which can be successfully cultivated. The likelihood of cold winters, warm summers and drought can be evaluated. The World Meteorological Organisation has set thirty years as the standard period for comparison of weather data on a world-wide basis, e.g. 1901–30, 1931–60 and 1961–90, as they believe that this is a long enough time to give an indication of the variability of the seasons. In Britain the 1961–90 period included the extremes provided by the severe winter of 1962–63 and the famous hot summer of 1976. There are also overlapping averages, 1941–70 being widely used.

Past weather patterns are a useful guide in forecasting weather. Also long-period records can help monitor climatic change, as changes between one thirty-year period and another are likely to be of the order of 0·1°C or 0·2°C (0·2°F or 0·4°F).

Architects, builders and civil engineers need to have some indication of the weather conditions which can be expected. How many working days will be lost, for example, by heavy rain? How big do the storm drains have to be to cope with a flood, which may be expected once every five years? How strong must the roof of a big stadium be to cope with a heavy snowfall? What are the maximum gusts expected on a bridge?

Water-supply companies need records to see where the rain falls regularly, so that reservoirs can be located in the right places; for annual rainfall can vary significantly over a few kilometres. Do droughts occur so often that huge and costly reservoirs are needed, or can the risk be taken that the inconvenience of the drought is less irksome than higher water bills?

Weather records are also used in calculating certain types of insurance risk, so insurance companies are very interested in weather data. In Britain premiums for storm-damage risks are lower in eastern England because adverse weather is less likely there. The cost of a pluvius policy for people running fetes and fund-raising activities is lower in eastern England than in the west. Quite a few amateur weather enthusiasts earn extra income by providing information to insurance companies about weather conditions in connection with road accident claims, as does the Meteorological Office at Bracknell. In other words 'they' check up on 'weather conditions at the time of the accident' and see whether what is written on the claim form fits.

While there would appear to be serious, or commercial, reasons for recording the weather, there are many people who keep a rain gauge or a maximum and minimum thermometer purely for interest and curiosity. The advent of electronic recording and of computers which will log and store the data will probably bring about an increase in their numbers, as there must be many people with irregular working hours who cannot record observations daily at the same time. Electronic equipment is much more sensitive to changes in pressure and temperature and this could pose a problem when comparing records in the future. Power cuts and nearby lightning strikes also have adverse effects on electronic equipment, and may erase records.

Rain has Method

The process of rainfall depends on the natural law that warm air is capable of holding more water vapour than cold air. Water vapour is the raw material of rain. For example, air at 20°C (68°F) holds 3·6 times more water vapour than air at 0°C (32°F). Table 4 shows the dewpoint and moisture content of air at different temperatures.

Relief has an important effect on rainfall, causing more rain to fall over the mountains than in the lowlands nearby (Fig. 28). Air can be cooled to its dewpoint by being forced to rise over a range of hills or mountains. The normal rate of cooling for unsaturated air is 1·0°C (1·8°F) for each 100 metres (328 feet) of ascent, and 0·6°C (1·1°F) after saturation. These rates of cooling are known as the dry adiabatic and saturated adiabatic lapse rates respectively, meaning that the drop in air pressure as the parcel of air rises is responsible for the drop in temperature of the rising air. Moist air cools more slowly because the process of condensation liberates latent heat to the rising parcel of air.

Saturated air at 20°C (68°F) holds 17·1 grams of water per cubic metre of air. If the relative humidity is 75%, it will hold 75% of 17·1 which is 12·8 grams per cubic metre, or the same amount as saturated air at 15°C (59°F) (Table 4(a)). If the air near the ground is cooled to this temperature, dew and then mist or fog will form, such as

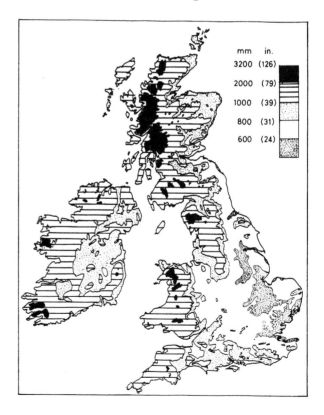

Fig. 28. **Average annual rainfall 1941–70**

might happen on a summer morning before sunrise. If the air was cooled by ascent instead, clouds would be likely to form at 500 metres (1640 feet). The difference between the surface temperature and its dewpoint (in °C) can be used to estimate the likely height of the cloud base in hundreds of metres. The formula works well for estimating the height of cumulus clouds, which form in well-mixed air in the middle of the day, but is not satisfactory in the early morning, especially when the air near the ground is saturated, or when a temperature inversion occurs in the middle atmosphere.

Table 4 **Moisture Content, Dewpoint and Relative Humidity of Air at Different Temperatures**

a. MAXIMUM WATER VAPOUR CONTENT OF AIR AT DIFFERENT TEMPERATURES

Air temperature (°C)	30	25	20	15	10	5	0	−5
Grams per cubic metre of air	30	23	17·1	12·8	9·4	6·8	4·8	3·3

b. DEWPOINT TABLE

Actual temperature (°C)	Relative humidity (per cent)																
	100	95	90	85	80	75	70	65	60	55	50	45	40	35	30	25	20
25	25	24	23	22	21	20	19	18	17	16	15	14	12	10	8	6	4
23	23	22	21	20	19	18	17	16	15	14	12	10	9	7	5	2	−1
21	21	20	19	18	17	16	15	14	13	12	10	9	7	5	3	0	
20	20	19	18	17	16	15	14	13	12	11	9	8	6	4	2	−1	
18	18	17	16	15	14	13	12	11	10	9	7	6	4	2	0	−2	
16	16	15	14	13	12	11	10	9	8	7	6	4	2	1	−1	−4	
14	14	13	12	12	11	10	9	8	6	5	4	2	1	−1	−3	−6	
12	12	11	10	10	9	8	7	6	5	3	2	1	−1	−3	−5	−7	
10	10	9	8	8	7	6	5	4	3	1	0	−1	−3	−5	−7		
8	8	7	7	6	5	4	3	2	1	0	−2	−3	−5				
6	6	5	4	4	3	2	1	0	−1	−2	−4	−5	−7				
4	4	3	3	2	1	0	−1	−2	−3								
2	2	1	1	0	−1												
0	0	−1	−2	−2													

Dewpoint temperature (°C)

The base of the lowest clouds, measured in hundreds of metres, is often the same as the difference in degrees Celsius between the surface temperature and its dewpoint: e.g. surface (screen) temperature 18°C, 75% relative humidity, dewpoint 13°C, difference 5°C, so that the cloud base is 500 metres.

Table 4 *continued*

c. RELATIVE HUMIDITY TABLE

Dry Bulb temperature (°C)	Depression of Wet Bulb (°C)									
	1	2	3	4	5	6	7	8	9	10
40	93	87	82	76	71	66	61	56	52	47
35	93	87	81	75	69	64	58	53	49	44
30	92	86	79	73	67	61	55	50	44	39
25	92	84	77	70	63	57	50	44	38	
20	91	83	74	66	59	51	44	37		
15	90	80	71	61	52	44	35			
10	88	76	65	54	44	34				
8	87	75	63	51	40					
6	86	73	60	47	35					
4	85	70	56	42						
2	84	68	52	37						
0	82	65	48							
−2	80	61	42							

Relative humidity (per cent)

The percentage relative humidity can be found by wet and dry bulb thermometers. The wet bulb thermometer is an ordinary thermometer with a piece of muslin tied around the bulb and with the other end immersed in distilled water which travels up the wick by capillary action. As evaporation cools the water, the wick in contact with the bulb of the thermometer will be cooler than the air, thus affecting the thermometer reading. In dry weather, the depression of the wet bulb will often be as much as 5° or 6°C (9° or 11°F) and on occasion as much as 10°C (18°F). In humid weather the two thermometers will read the same. Example, dry bulb 20°C (68°F), wet bulb 15°C (59°F): percentage humidity 61%.

At 80% humidity and a temperature of 10°C (50°F), far less cooling is needed to bring the air to its dewpoint. At these values the dewpoint is 7°C (45°F) (Table 4(b)) and clouds would be expected to cover hills above 300 metres (1000 feet)), shrouding Dartmoor and the fells of the Lake District, for example. Even moister air at 15°C (59°F) with 90% relative humidity, as is likely to be found in the warm sector of a depression, has a dewpoint of 13°C (55°F); in such circumstances clouds would be expected above 200 metres covering all but our lowest hills.

Rain brought about by an air current being forced to rise over a range of hills is known as 'orographic' or 'relief' rain, that is, it is due to the relief. Even small hills have a significant effect. At Cockpit Hill (152 metres, or 499 feet), the average annual rainfall is 187 mm (7.4 in) more than in Nottingham, only 9 kilometres (6 miles) to the south. The extra can be considered as orographic rainfall. Further evidence from Surrey shows not only that rainfall is heavier over the hills but that it occurs more often, a fact which is well known for Snowdonia or Dartmoor, but is not so widely appreciated in the undulating lowlands. Cranleigh, Surrey, at an altitude of 47 metres (154 feet) has rain on 167 days in the year on average; Peaslake, 81 metres (266 feet) higher, 11 kilometres (7 miles) away, has an annual average of thirty-one more days a year, or over two more per month, and yearly rainfall total 117 mm (4.6 in) higher than at Cranleigh, brought about by the difference in altitude.

Air will also be cooled to its dewpoint when a warm current of air, such as Tropical Maritime, comes into contact with Polar Maritime. As the tropical air is warmer, it is lighter and will tend to rise above the colder polar air and be cooled adiabatically. Some hours of steady rain will be most probable, falling from a thick layer of nimbo-stratus and accompanied by a rise in temperature at ground level. This is warm-front rain and in the lowlands of the east it is not usually very heavy; however, in the warm sector behind the front, large amounts of rain can sometimes be expected over the moors of the west and north.

A shower is convection rain. Showers are likely when cold air has a long crossing over a warmer sea surface, or when strong sunshine heats up the ground sufficiently to cause warm bubbles of air to rise vertically and form shower clouds. Ideal conditions for convective clouds to form over land are: relatively low pressure, a large lapse rate (or fall of temperature with height), and a supply of moist air. For if the air is dry, even though it may be lifted to a great height it may not be cooled to its dewpoint.

Sometimes orographic rain falls only over the hills while the lowlands remain dry. But orographic and frontal rain often occur at the same time. As a front passes, more rain falls over the moors than in the lowlands. Rain will also be likely near the cold front (Fig. 20) when the cold air in the rear of a depression undercuts the warmer tropical air, lifting the milder air more vertically here than at the warm front, so the rain is often heavier. When the cold air behind the depression moves faster than the warm front and catches up, the air in the warm sector is entirely lifted off the ground (or occluded) (Fig. 21), a situation which can produce some very heavy downpours. It is a sign that a depression is decaying, or filling, when the warm sector is occluded. The occlusion of the air in the warm sector has a very similar effect to the passing of that air over a range of hills or mountains.

Measuring Rainfall

The amount of rain is measured by a gauge of simple but accurate design. A circular funnel, usually 127 mm (5 in) in diameter, catches the rain and passes it to an inner can. Every morning, usually at 9 h Greenwich Mean Time, the water in the container is emptied into a specially graduated measuring glass. The smallest amount measured is 0·1 mm, which is about one two-hundred-and-fiftieth of an inch. Before 1961, when the Meteorological Office began to go metric, measurements were taken to the nearest 0·01 in. If the fall of rain is less than 0·1 mm the word 'trace' is entered in the register. A standard and recording gauge are shown in Fig. 29. The trace from a recording gauge is shown in Fig. 30. Anyone can make a simple rain gauge by placing a funnel in a bottle; but make sure that the rain cannot run down the outside of the funnel and enter the bottle that way. To calibrate a measuring glass, first calculate the area of the circular mouth of the funnel using the formula πr^2, where π is 3·1416 and r is the radius of the funnel. The catchment area of the funnel is 126·7 cm^2. Then 1 cm or 10 mm of rainfall would yield 126·7 cm^3. If this much water were poured into a measuring glass, equal divisions could then be scratched on the surface so that the amount could be read off the scale, i.e. the water level would represent 10 mm of rain. If fifty equal divisions were made, then the gauge could be read to the nearest 0·2 mm.

Fig. 29. **The rainfall station at Seathwaite, Borrowdale, Lake District.** This is probably Britain's wettest inhabited place with an average annual rainfall of 3120 mm (122·8 in). On the *left* is the recording gauge with a revolving drum, driven by electricity, on which a pen arm with a special nib draws a line to record both the amount of rain and the time when rain was actually falling. In the *centre* is a standard gauge used as a check in case of mechanical trouble. To the *right* is a gauge designed for some isolated places, which can be read as infrequently as once a month. Here its performance can be measured against the standard daily instrument. In the *background* are the fells around Sprinkling Tarn, where a monthly gauge is in use, which vies with Snowdon for the highest annual totals in the British Isles.

Average monthly rainfall at Seathwaite (1941–70) in mm.

Jan	Feb	Mar	Apr	May	Jun	Jul	Aug	Sep	Oct	Nov	Dec	Year
313	325	199	199	168	187	239	273	343	318	323	333	3120

Fig. 30. *(below)* **Rainfall at Guildford on 10 September 1976 (from a recording rain gauge).** The numbers along the top of the chart indicate the time by the twenty-four-hour clock. The figures by the horizontal straight lines show millimetres of rain. The pen draws a straight line when no rain is falling: the steeper the slope of the line on the chart, the heavier the rainfall. Only 5 mm of rain fell between 17·30 and 22·30 h, the pen arm then returning to zero. The next 5 mm fell in just under one and a half hours. The rain continued till 05·30, except for a short break between 02·30 and 3·00. From 03·45 until just after 5·00, about 30 mm of rain was recorded, a very heavy downpour. The total for the rainfall day was 50·5 mm (1·9 in).

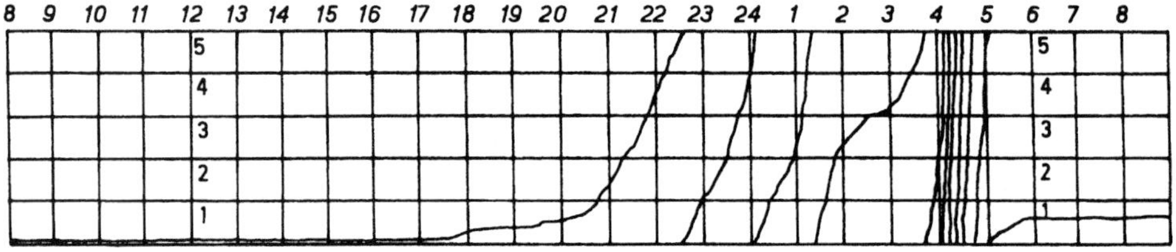

Probably the easiest and cheapest method for most of us of acquiring a rain gauge is to use a standard 127 mm (5 in) funnel, from a hardware store, and to buy the measuring glass from any of the well-known meteorological suppliers whose names are given on page 292. Almost any container will do, provided it is watertight. However, beware of glass! It will disintegrate in frost. The gadget-minded can rig up a central-heating system in the form of an electric light bulb, essential for recording gauges. Electronic gauges, which are often supplied with 10 metres of cable, have a display unit which can be installed within the house, so morning trips into the garden can be avoided, but like all things electronic they are fine when they are working well and really do need daily checking. A slug or earwig on the mechanism may cause incorrect readings.

Enthusiasts who would like to keep records often find that buildings or tall trees give an inaccurate result. The Meteorological Office lays down exacting requirements for the siting of gauges. The distance away from every surrounding object should be not less than twice the height of that object above the rim of the gauge and preferably four times the height.

Distribution of Rain

Fig. 28 shows that most rain falls over the mountains and moors of the west where some places have as much in a year as falls on the Fens in eight years. Styhead, near Scafell in the Lake District, has the largest average annual total so far measured in the British Isles: 4306 mm (169·5 in). Crib Goch in Snowdonia has only a fraction less with 4282 mm (168·6 in). Although the largest amount of territory with over 2500 mm (98 in) lies in Scotland, competition for the national record for wettest place seems to lie between the Lakes and Snowdonia. The two wettest places in Scotland are Coire nan Gall with an annual average of 4193 mm (165·1 in) and Cruadhach with 4127 mm (162·5 in).

The east coast from Yorkshire to Kent, the Fens and much of East Anglia have least rain. Small areas in Scotland around the Firth of Forth and Moray Firth are nearly as dry, for they are low-lying and sheltered from the rain-bearing winds by high ground. St. Osyth, near Clacton-on-Sea, Essex, has the lowest annual average in the country of 513 mm (20·2 in) for the period 1964–82.[2] Over thirty-five years to 1960 Great Wakering, near Southend, had an average of 487 mm (19·2 in) while Bottisham Lock, Cambridgeshire, had an average of 512 mm (20·2 in) for the period 1931–60.

Few of us know what great downpours can occur over the mountains. Seathwaite (Fig. 29) with an annual average of 3120 mm (122·8 in) is probably the wettest inhabited place in England. Many Londoners would be staggered if their rainfall increased fivefold to the Seathwaite total; even more if it approached the volume of water which falls on Snowdon or Sprinkling Tarn. Over the lowlands 50 mm (2 in) a month is a common standard, giving an annual total of 600 mm (24 in) . In the west 900 to 1000 mm (35 to 39 in) is general.

Moist air moving from the Atlantic, for example, does not have an infinite supply of water vapour. If some is deposited as rain over the coasts and hills of the west, then places in the east will receive less. Just as air can be cooled to its dewpoint by forced passage over a range of hills, so the reverse process can take place when the air descends the lee slope. It then becomes warmer and drier and more able to hold what moisture is left in the air. This partly explains why the east has less rain than the west, Yorkshire has less rain than Lancashire and Inverness less than Skye. Places sheltered from rain-bearing winds by high ground (such as Teesside and Lincolnshire) are said to be in a rain-shadow area.

If a line is drawn from Essex to South Wales, we find that Shoeburyness has an annual fall of 539 mm (21·2 in), Kew 599 mm (23·6 in), Reading 668 mm (26·3 in), Marlborough 822 mm (32·4 in), Cardiff 1059 mm (41·7 in) and Swansea 1158 mm (45·6 in). None of these places is on high ground. Thus the most westerly has over twice as much rain as the most easterly. Ignoring the influence of physical features, then, rainfall increases from east to west, which is not surprising as most of our rain comes from the Atlantic. At Keele, near Stoke-on-Trent, frontal rain, associated with depressions mostly from the Atlantic, accounts for about 60% of rainfall. Showers mainly associated with Polar Maritime air account for a further 19%.[3] Information collected by the Building Research Centre[4] provides extra evidence and shows that winds from the Atlantic (taking winds between north-west and south) provide two-thirds of the rainfall (Table 5).

And what of Manchester? This city is notorious for rain. Is it a fallacy? The average annual rainfall at Ringway Airport nearby is 819 mm (32·2 in). Compare this with the averages for other places in the district. Oldham averages 1142 mm (45 in), Rochdale 1130 mm (44·5 in), Liverpool 849 mm (33·4 in) and Southport 836 mm (32·9 in). No one talks of continuous rain there. Bournemouth also has the same rainfall as Manchester, and famous resorts like Falmouth, Penzance and Ilfracombe have more rain than Manchester. Statistics show that Manchester has rain on about 195 days a year on average. So do Southport and Falmouth.

Table 5 The Proportion of Rain Falling on a Vertical Aperture on a Building, Facing the Directions Shown (north is taken as 1·0)

N	NE	E	SE	S	SW	W	NW
1·0	1·13	1·27	2·25	3·5	4·0	2·8	1·37

Table 6 Summary of Whole Days Lost Due to Rain in Test Match Cricket: 1880–1996

Ground	Whole Days Lost	Test Matches Scheduled
Lord's	17	95
The Oval	13	79
Old Trafford	28	64
Headingley	12	58
Trent Bridge	9	44
Edgbaston	3	32
Sheffield	0	1

Summary of Saturday test days completely washed-out and of Saturdays with less than one hour of play (in brackets).

Lord's	5 (1)
Old Trafford	4 (3)
The Oval	3 (1)
Headingley	2
Trent Bridge	1 (1)

The traditional raininess of Manchester is a myth. Indeed, Manchester's rainfall is an average one for England as a whole, about 830 mm (32·7 in). How has the legend arisen? Is it the number of test matches at Old Trafford interrupted by rain? There is certainly something in that, as the statistics (Table 6) show; for the chance of a wholly wet day interrupting an Old Trafford test is just under one match in two compared to one in six at Lord's or The Oval.[5] Also, Old Trafford has the unenviable distinction of experiencing the only two wholly washed-out tests: in 1880 and 1938. Perhaps cricket, then, or rather the reporting and televising of it, has helped the myth of rainy Manchester. Or is it that of all the test grounds in the country Old Trafford is the most prone to rain as it is the most westerly? In fairness to the England and Wales Cricket Board, it should be acknowledged that comparatively few days have been completely lost since 1983, which is probably a reflection of more efficient protection of the wicket, as well as prevailing weather conditions.

Raininess

What are the greatest and least amounts of rainfall that have been recorded in the British Isles in a year? As much as 6000 mm (236 in) were measured in 1872, 1898, 1909, 1923 and 1954. In 1923 as much as 6283 mm (247·4 in) fell at The Stye, Borrowdale; but the record goes to Sprinkling Tarn, near Scafell, when 6527 mm (257 in) fell in 1954. At the other extreme Margate totalled 236 mm (9·29 in) in 1921. This is the only occasion when less than 254 mm (10 in) has been recorded anywhere in the British Isles in a calendar year, or probably in any consecutive twelve months. If such conditions were to persist, the country would soon become a desert. Other low annuals include 332 mm (13·1 in) at Austerfield, West Yorkshire, in 1959; 338 mm (13·3 in) at Geanies House, Highland (Ross and Cromarty), in 1972. In Wales, St. Asaph recorded 478 mm (18·8 in) in 1955. In Northern Ireland, Donaghadee registered 504 mm (19·8 in) in 1952. In the twelve months to 31 August 1976 only 318 mm (12·5 in) of rain fell at Newark and 315 mm (12·4 in) at Royston, Hertfordshire. But drought deserves a chapter to itself (Chapter 10).

Fig. 31.   **Ninety-five years of rainfall, 1900–94.** Each dot represents one year's rainfall.

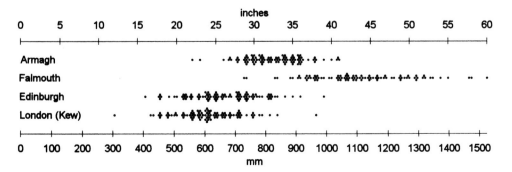

In every year from 1911 to 1994, some locality reported at least 3000 mm (118 in), and during the same period a fall of less than 508 mm (20 in) was noted at some place in Britain in three-quarters of the years. Some idea of the annual variation of rainfall can be gained from Fig. 31, which shows annual rainfall this century at Armagh, Falmouth, Edinburgh (Blackford Hill) and London (Kew). Only one year this century (1933) has departed greatly from the average at Armagh. Another notable feature is that the wettest year at Kew is not far below that for Armagh.

Table 7 shows the 1941–70 averages for the places shown in the scatter graph, and for the wettest and driest years.[6] Variability appears to be least at Armagh and greatest at Falmouth. Also, the wettest years were different at the four stations; so were the driest.

Table 7 **Average Rainfall and Extreme Annual Totals**

	Average (1941–70)		Wettest Year			Driest Year		
	(mm)	(in)		(mm)	(in)		(mm)	(in)
Armagh	864	34·01	1958	1039	40·9	1933	548	21·59
Edinburgh	699	27·51	1916	990	38·99	1902	418	16·44
Falmouth	1149	45·23	1924	1480	58·29	1953	726	28·58
Kew	599	23·58	1903	970	38·18	1921	309	12·16

So far the raininess of a place has been described in the traditional way, using average rainfall totals, but it is probable that frequency of rain makes a greater impression on our minds than amount. A day with rain, or a 'rain day' in official circles, is defined as one in which the fall is 0·2 mm (0·01 in) or more. Many rain days have trifling amounts. Most of England has fewer than 200 a year.[7] The south, the Midlands and much of eastern England south of the Wash have less than 175 days with rain. Localities with over 200 include the Peak District of Derbyshire, the Lake District, most of Scotland except the east coast, the greater part of Wales and Ireland and the moors of the south-west. Small areas in Scotland have over 250 rain days a year. Stornoway averages 260, as does Lerwick. At Baltasound, in Shetland, rain occurred on an average of 296 days a year over a ten-year period. Most people in England would object if a single month had more than twenty rain days. In individual years a place or two will record over 300 rain days. In 1923 Ballynahinch, Co. Galway, recorded 309 rain days. A ten-year average for 1985–94 for Galway City shows that rain is likely on 233 days.[8] Only four months had less than ten days with rain: February 1986 with one; October 1993 with eight; June 1986 with nine and September 1986 with nine.

It is rare for less than 100 rain days to occur anywhere in Britain. In 1921 Chatteris, Cambridgeshire, had only ninety-three and Birchington, Kent, only eighty-four. In 1949 several places around the Thames estuary had 111 rain days. In the twelve months to the end of August 1976 (when the great drought ended) rain fell on 109 days in several parts of the south.

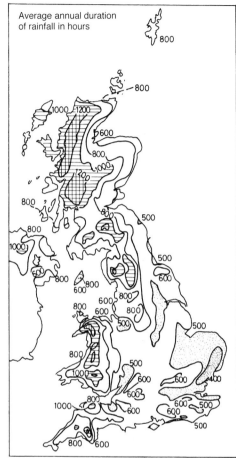

Fig. 32. (*above*) A rain day is a day on which 0·2 mm or more of rain (or equivalent in melted snow or hail) is recorded in the gauge

Fig. 33. (*right*) Information from recording gauges has been used to compile this map

Although the rainfall in our wettest spots may be from six to eight times that in our driest, the difference in the number of rain days is much smaller. The number of 'wet days' (when 1 mm (0·04 in) or more of rain falls), gives a better picture of raininess, even though it still does not mean what most people would call a wet day. This much rain could easily fall in a summer shower, lasting ten minutes or so, an amount which could make the unprepared quite wet unless shelter was at hand. The ratio of wet days to rain days is about two to three in the south-east and Midlands, but rises to 90% in the mountainous parts of the country: that is, from 110 to 240 days a year. 'Very wet days' (with rainfall exceeding 10 mm (0·4 in)) vary from under fifteen around the Thames estuary and the Fens to 110 in the Highlands. At Kew heavier downpours are fairly rare and 25 mm (1 in) or more in a day occurs about once a year on average. Over the twenty years 1951–70 more than 25 mm occurred four times in 1964 and three times in 1956. Besides total rainfall and the number of rain days, there is another method of assessing the raininess of a place. Thanks to recording rain gauges we are able to produce maps (Fig. 33) which show rainfall duration in hours. Compare the rainfall at Kew Observatory (where the average is 599 mm (23·6 in) falling on 153 days a year) with Glenleven, Strathclyde (Argyll), which receives 2100 mm (83 in) and has 266 days when precipitation is recorded. This gives a daily average of 3·9 mm (0·15 in) at Kew and 7·9 mm (0·31 in) at Glenleven, and it is tempting to infer that it rains more than twice as heavily in the Highlands as it does in the London area.

The inference is incorrect since it leaves out the question of how much of each day it is actually raining. If rainfall is measured by duration, the very wet districts may have rain for six times as many hours as the driest. Blackwater Dam, Glenleven, has an annual average of 1560 hours per year. That is more than four and a quarter hours a day, compared to one hour and twelve minutes at Kew. In 1949 parts of the Essex coast had 264 hours of rain and Loch Sloy 1678.

If the average rates of fall at Kew and Glenleven are compared, then the rate of fall appears to be about the same, or 1·4 mm (0·06 in) per hour. This may be surprising and provoke the protest, 'but it simply pours in the Highlands.' So it does sometimes. However, the heavy mountain downpours are more notable for persistence than great intensity. Over most of the British Isles the average rate of fall is about 1 mm (0·04 in) per hour. Cwm Dyli at the foot of Snowdon appears to have about twice this intensity. Two points should be remembered. In mountainous districts, frequent drizzle reduces the average intensity of the rainfall; in the lowlands, summer thunderstorms would increase the average. It is reasonable to assume that on wet days in the Highlands the rain falls with considerable intensity, though it is not as intense as a lowland thunderstorm.

NOTES TO CHAPTER 4

1. Met.O. 856c, Part III, Europe and the Azores, HMSO.
2. M.O. and *Guinness Book of Records 1996*, Guinness Publishing, 1995.
3. S.H. Beaver and E.M. Shaw, *The Climate of Keele*, Keele University Library, 1970, p. 69.
4. Building Research Centre, Garston, Watford.
5. Irving Rosenwater, 'The lost days of Test Match cricket', *Cricket Quarterly*, Curator, MCC, Lord's Ground, London. *Wisden 1975–8*.
6. M.W.R. (Annual Summary) and M.O.
7. J. Glasspoole, 'The distribution over the British Isles in time and space of the annual number of days with rain', B.R. 1926, pp. 260–79; A.B. Tinn, *This Weather of Ours*, Allen and Unwin, 1949.
8. C.O.L. and F. Gaffney, Galway.

5 | What Rain Can Do in a Month

Tropical climates have wet and dry seasons. Bombay, for example, has an annual average rainfall of 1809 mm (71·2 in); but four months there are virtually rainless, while in July as much rain falls as the driest parts of England receive in a year.[1] Around the Mediterranean there are wet and dry seasons which are normally fairly reliable for the type of weather they bring. In Britain, by contrast, rainfall is spread quite evenly over the year, though there are some months when the fall is relatively heavy and others when it is comparatively light. Generally spring is the driest season; however, 'What about April showers?', you may say. Yes, in spite of April showers, spring is the driest season on average, as Table 8 shows,[2] and April is the driest month at just under half of the places listed in the table, followed by March, June, May and February in that order.

November or December between them are the wettest months at about two-thirds of the places shown in the table, which is to be expected as this is the time of the year when many Atlantic depressions come near to the British Isles. Of the remaining places, August has the most rain on average.

Information from automatic recording gauges shows that the duration of rain is least in April, May, June and July (Table 9).[3] August figures quite prominently as the wettest month at a number of places, yet it is not a top month for hours with rain anywhere. This suggests that August has a tendency to produce heavy thundery rain which does not last long. At the majority of stations shown in the table, December is the month with the highest duration totals, followed by January, with November a long way behind.

At Kew, April and July are the driest months with an average duration of fifty-four minutes a day. January and December are wettest with rain lasting for about an hour and a half a day on average. In Edinburgh, the daily rainfall duration is just an hour in April and over two hours in July, August and September. May is the driest month in the Highlands and at Loch Sloy the monthly average is seventy-seven hours or about two and a half hours a day, while the December total is 207 hours or nearly seven hours of wet weather each day, though it must not be forgotten that this includes frequent mountain drizzle as well as downpours. Table 11 summarises the information about rainfall for Kew, Plymouth, Manchester (Ringway) and Eskdalemuir.[4] It will be observed that Plymouth is wetter than Kew in all months, but notably so in the winter; from April to September the difference is less. Thus the fear of rain need not deter anyone from taking a holiday in Devon, although it is true that Plymouth has more rain than London, however we measure it.

In individual months great differences from the average may occur. An exceptionally wet month may have two, three or even four times the average rainfall; an unusually dry one nothing at all. At Plymouth the variation from the average in June is from 217% to 0%; at Kew from 389% to 2%. Anyone who cares to examine rainfall records will discover that it is uncommon for a year to pass without some place in Britain experiencing a monthly fall of 625 mm (25 in) or the equivalent of London's fall in twelve months. A month with 625 mm is very difficult to imagine, because a lowland fall of 125 mm (5 in) would be considered very wet. That leads us to the question, 'What is the greatest rainfall recorded in a single month in Britain?' In October 1909, Llyn Llydaw, Snowdon, at a height of 457 metres (1499 feet) above sea-level, registered 1435 mm (56·5 in) or more than 46 mm (1·8 in) a day, approaching a rate of fall of 2 mm (0·08 in) per hour. On three other occasions a fall of 1270 mm (50 in) has been recorded, in January 1872 and 1928 in Borrowdale, and in March 1938 at Loan, Highland (Inverness-shire). Incredible as it may seem, March 1938 was one of the driest on record over much of England and eastern Scotland. How many people living in that dry March in the English lowlands would have thought that a few hundred kilometres away as much rain was falling as they would normally receive in two years? But there was an even more remarkable contrast in Scotland itself. While 1271 mm (50·04 in) was falling on Loan, and 1017 mm (40·03 in) at Kinlochquoich, Highland (Inverness), a mere 225 kilometres (140 miles) away at East Fortune, near Dunbar, only 5 mm (0·2 in) was recorded. An anticyclone over southern England was giving fine weather there, while moist westerly winds were bringing Tropical Maritime air to the north-west with copious rain, both relief and frontal.

Our rainfall varies with the seasons. But as a rule it is well distributed throughout the year. Sometimes, though, this normal distribution is quite blotted out. So great is the variation that it is possible for any month to be the wettest or driest of the year over most of the country. In the Midland city of Nottingham, records over seventy-six years show that July, August and October have each been the wettest (i.e. recorded the largest monthly total in a year) on thirteen occasions. At the other end of the scale, January has been the wettest month three times, March twice and February and April once each.

Some Exceptional Months [5]

August 1912

The dreary summer of 1912, which came directly after the hot summer of 1911, brought one of the wettest Augusts in meteorological annals. Cornwall and Norfolk received more than three times the average rainfall. It was the month of the great Norfolk rainstorm which will be described in the next chapter.

Table 8 **Monthly Rainfall Averages (in mm), 1941–70**. Selected stations shown in italics refer to 1961–90.

Height (metres) above sea level		Jan.	Feb.	Mar.	Apr.	May.	Jun.	Jul.	Aug.	Sep.	Oct.	Nov.	Dec.	Year (mm)
	SCOTLAND													
82	Lerwick	130	101	85	79	58	62	67	72	111	133	130	144	1172
	Lerwick	*133*	*90*	*113*	*72*	*62*	*60*	*62*	*78*	*118*	*133*	*143*	*139*	*1204*
15	Stornoway	111	84	76	72	57	66	75	83	106	122	112	130	1094
	Stornoway	*122*	*84*	*105*	*64*	*60*	*63*	*73*	*83*	*116*	*138*	*133*	*125*	*1167*
36	Wick	82	60	53	44	47	50	63	76	70	74	83	86	788
5	Kinloss	84	39	34	36	49	53	68	88	55	59	56	55	676
58	Aberdeen (Dyce)	78	56	50	47	68	56	80	84	67	80	86	75	827
	Aberdeen (Dyce)	*82*	*52*	*58*	*54*	*60*	*54*	*61*	*73*	*68*	*78*	*75*	*73*	*788*
10	Leuchars	57	42	40	39	57	46	67	75	58	58	65	57	661
134	Edinburgh (Blackford Hill)	54	41	36	38	58	47	75	86	63	57	64	54	673
335	Blackwater Dam	241	204	178	172	141	150	158	188	245	288	236	301	2502
107	Glasgow (Springburn)	93	71	63	63	74	73	87	104	112	107	95	102	1044
9	Tiree	115	74	70	65	58	70	85	88	127	133	114	130	1129
	Tiree	*127*	*79*	*96*	*59*	*59*	*61*	*78*	*94*	*129*	*140*	*122*	*120*	*1163*
	ENGLAND													
72	Acklington	61	48	40	39	48	42	61	76	57	52	70	50	644
78	Newcastle	61	50	38	41	52	52	62	82	60	53	72	53	676
52	Scarborough	60	48	38	42	48	46	56	66	52	53	72	59	640
9	York	55	44	37	42	46	44	53	75	55	47	63	49	610
9	Doncaster	50	43	39	40	47	47	57	71	49	45	63	43	594
80	Wittering	50	40	41	40	51	49	53	65	48	57	59	47	600
28	Norwich	57	49	43	41	44	44	62	67	54	57	68	57	643
43	Ipswich	54	41	41	39	40	48	56	61	53	55	66	56	610
6	Dungeness	75	50	41	39	42	38	53	61	60	71	88	67	685
120	Boscombe Down	72	46	48	45	57	47	47	67	64	74	82	75	724
32	Exeter	83	58	55	47	61	41	51	68	67	72	90	81	774
36	Plymouth Hoe	106	74	75	56	67	55	69	83	83	89	107	107	971
51	Scilly	92	77	69	47	65	49	62	71	72	78	99	100	881
9	Weston super Mare	72	50	49	49	65	50	63	85	78	76	90	79	806

Table 8 continued **Monthly Rainfall Averages (in mm), 1941–70.** Selected stations shown in italics refer to 1961–90.

	Height (metres) above sea level	Jan.	Feb.	Mar.	Apr.	May.	Jun.	Jul.	Aug.	Sep.	Oct.	Nov.	Dec.	Year (mm)
ENGLAND continued														
Oxford (Radcliffe)	63	59	41	43	42	56	50	59	68	60	58	63	60	659
Birmingham (Edgbaston)	37	69	55	56	55	71	55	68	80	67	64	81	73	794
Shawbury (Shropshire)	72	56	43	43	43	60	49	62	72	60	57	65	60	670
Liverpool	39	73	53	48	52	63	56	73	90	80	77	90	77	832
Blackpool	20	79	56	48	56	61	62	78	99	97	91	90	88	905
Carlisle	26	72	51	47	54	59	64	83	102	97	85	82	75	871
ISLE OF MAN														
Ronaldsway	16	90	60	59	53	56	52	58	77	85	89	92	95	866
WALES														
Valley (Anglesey)	10	90	60	51	52	58	55	61	77	89	87	97	94	871
		83	*56*	*65*	*53*	*49*	*52*	*53*	*74*	*73*	*91*	*101*	*93*	*842*
Llandudno	4	78	56	50	48	49	44	47	68	72	78	83	83	756
Aberystwyth	4	84	61	60	63	62	68	85	100	99	101	95	103	982
Swansea	10	116	76	76	70	77	71	85	105	112	113	128	129	1158
Crib Goch (Snowdon)	713	462	322	280	270	235	273	336	376	428	422	419	459	4282
NORTHERN IRELAND														
Aldergrove	68	86	59	55	54	60	67	79	88	88	86	84	94	900
		86	*58*	*67*	*53*	*60*	*63*	*64*	*80*	*84*	*87*	*78*	*77*	*858*
Armagh	62	78	57	53	58	61	68	75	87	80	83	77	87	864
CHANNEL ISLES														
Jersey	9	95	65	53	52	52	38	50	69	73	78	111	109	845
REPUBLIC OF IRELAND														
Cork	153	141	113	93	64	78	63	59	81	90	119	103	129	1132
Shannon	14	98	72	71	56	60	63	57	79	82	93	95	99	924
Valentia	9	167	121	122	77	88	80	73	106	124	157	148	159	1422

Table 9 **Average Monthly Hours of Rain, 1951–60**

	Jan.	Feb.	Mar.	Apr.	May.	Jun.	Jul.	Aug.	Sep.	Oct.	Nov.	Dec.	Year (hours)
Lerwick	99	76	57	56	50	51	59	56	60	78	92	106	840
Stornoway	105	62	58	60	52	55	59	57	56	85	84	101	833
Wick	80	62	55	42	44	48	53	57	44	62	72	76	695
Tiree	102	70	44	58	46	58	68	62	71	89	87	116	871
Aberdeen (Dyce)	58	57	50	43	40	39	61	51	39	54	61	69	621
Loch Sloy	144	107	105	95	77	88	104	108	105	139	143	207	1422
Edinburgh (Turnhouse)	49	47	44	34	44	51	64	62	42	43	57	70	606
Acklington (Northumberland)	61	48	46	34	40	45	41	56	39	46	54	54	564
Carlisle	65	45	41	42	43	49	56	61	58	57	74	77	668
Manchester (Ringway)	64	48	48	41	34	47	52	55	50	54	58	71	623
Shawbury (Shropshire)	61	47	47	31	40	47	36	48	41	46	58	58	559
Birmingham (Elmdon)	59	47	56	35	38	40	34	45	41	47	62	61	566
London (Kew)	45	42	36	27	30	28	27	31	33	40	43	45	427
Folkestone	66	59	42	30	29	29	31	35	35	49	57	65	527
Exeter	62	52	51	35	41	30	28	35	39	49	64	67	553
St. Mawgan (Cornwall)	78	58	57	38	41	34	36	44	50	53	69	80	637
Valley (Anglesey)	75	61	53	41	49	44	46	52	59	61	80	81	703
Aberporth (Dyfed)	84	56	69	45	53	51	49	56	65	73	91	90	781
Swansea	75	51	58	45	45	43	48	52	58	64	78	82	699
Jersey (1951–60)	66	56	43	31	29	24	30	35	38	43	48	72	515
Jersey (1961–91)	72	61	56	41	38	27	23	23	36	46	67	74	565
Ronaldsway (Isle of Man)	61	50	41	34	39	40	40	44	54	55	67	74	599
Aldergrove	80	59	51	45	44	55	59	62	53	54	66	95	723

Table 10 **Average Monthly Hours of Rain, 1985–94**

	Jan.	Feb.	Mar.	Apr.	May.	Jun.	Jul.	Aug.	Sep.	Oct.	Nov.	Dec.	Year (hours)
Denbury (Devon)	100	83	76	58	40	39	42	48	46	69	79	92	772
Hove (West Sussex)	64	37	49	44	29	29	31	28	32	46	53	52	494
Moel-y-Crio (Clwyd)	72	69	81	74	51	61	50	64	50	85	78	95	830
Galway (Western Ireland)	104	75	109	85	55	68	66	82	66	84	79	114	987

Table 11 **Rainfall Data**

KEW OBSERVATORY (5 m), 1866–1980 (station closed), *HEATHROW 1961–90 average monthly falls in italics*

	Average monthly fall (mm)		Average number of rain days	Maximum fall in 24 hours (mm)	Highest monthly fall (mm)	Year	Lowest monthly fall (mm)	Year	Average duration (hours)	Highest duration	Lowest duration 1941–70
January	49	*52*	15	29	127	1877	11	1880, 1892	49	82	20
February	38	*35*	13	22	127	1951	2	1895, 1959	40	104	6
March	38	*47*	11	28	118	1947	1	1929	38	105	5
April	40	*45*	12	24	99	1878	4	1912	36	85	7
May	48	*51*	12	25	104	1886	4	1896	34	73	9
June	47	*50*	11	36	183	1903	1	1925	28	57	7
July	58	*46*	12	60	151	1956	4	1921	27	53	3
August	63	*51*	11	56	165	1878	2	1940	32	61	10
September	52	*51*	13	46	145	1918	3	1959	33	63	9
October	52	*58*	13	35	151	1880	2	1978	45	61	4
November	62	*55*	15	35	172	1940	7	1945	50	104	9
December	52	*57*	15	26	162	1914	6	1926	47	87	15
Year	599	*598*	153						459	574	298

PLYMOUTH (MOUNT BATTEN) (27 m), 1921–81, *1961–90 average monthly falls in italics*

	Average monthly fall (mm)		Average number of rain days	Maximum fall in 24 hours (mm)	Highest monthly fall (mm)	Year	Lowest monthly fall (mm)	Year	Average duration (hours)	Highest duration	Lowest duration 1941–70
January	108	*114*	19	38	215	1984	13	1987	78	134	22
February	75	*92*	15	36	205	1977	1	1932	55	135	2
March	74	*87*	14	38	196	1977	3	1944	55	133	5
April	57	*59*	12	25	153	1961	4	1976	43	102	10
May	68	*61*	12	39	127	1942	6	1989	44	80	13
June	56	*57*	12	35	145	1968	0	1925	34	58	7
July	70	*55*	14	68	152	1936	7	1934	38	70	11
August	85	*69*	14	77	159	1952	2	1940	43	71	2
September	86	*76*	15	43	181	1974	0	1959	47	91	0
October	92	*95*	16	42	210	1960	7	1978	53	116	15
November	109	*101*	17	52	274	1929	25	1945	71	124	26
December	110	*116*	18	43	246	1934	14	1926	74	158	23
Year	990	*982*	178						634	794	417

Highest and lowest monthly falls correct to 1990

Table 11 *continued* **Rainfall Data**

MANCHESTER, RINGWAY (75 m), 1942–81, 1961–90 *average monthly falls in italics*

	Average monthly fall (mm)		Average number of rain days	Maximum fall in 24 hours (mm)	Highest monthly fall (mm)	Year	Lowest monthly fall (mm)	Year	Average duration (hours)	Highest duration	Lowest duration 1941–70
January	71	*69*	21	27	170	1948	11	1963	64	101	22
February	53	*50*	15	37	161	1977	4	1986	47	119	12
March	48	*61*	15	16	125	1988	13	1944	49	127	22
April	53	*51*	13	35	92	1970	4	1974, 1980	51	107	15
May	60	*61*	16	25	257	1986	11	1970	43	77	11
June	61	*67*	14	20	137	1958	12	1942	46	90	18
July	76	*65*	14	65	187	1973	13	1984	48	76	10
August	91	*79*	13	96	201	1956	7	1976	53	135	10
September	76	*74*	15	33	171	1965	5	1959	50	132	3
October	75	*77*	15	38	139	1954	14	1946	53	109	15
November	78	*78*	19	26	171	1951	11	1945	63	141	23
December	77	*78*	17	27	180	1965	9	1963	75	114	9
Year	819	*810*	187						639	895	433

ESKDALEMUIR (242 m), 1911–81, 1961–90 *average monthly falls in italics*

	Average monthly fall (mm)		Average number of rain days	Maximum fall in 24 hours (mm)	Highest monthly fall (mm)	Year	Lowest monthly fall (mm)	Year	Average duration (hours)	Highest duration	Lowest duration 1941–70
January	155	*166*	22	56	390	1928	43	1941	124	183	66
February	106	*110*	18	45	345	1990	5	1932	97	145	41
March	100	*137*	17	44	282	1989	17	1944	93	152	18
April	100	*83*	17	42	254	1947	17	1980	88	155	61
May	99	*95*	15	43	236	1925	4	1984	79	169	24
June	102	*94*	17	106	260	1931	22	1988	81	134	38
July	114	*98*	19	57	269	1988	25	1913	85	158	26
August	133	*125*	18	54	345	1985	1	1947	91	138	1
September	151	*149*	18	58	308	1950	18	1972	100	197	37
October	144	*164*	20	61	331	1967	12	1951	104	170	31
November	147	*152*	21	64	297	1938	18	1945	116	205	24
December	155	*165*	22	65	339	1932	33	1933	125	234	62
Year	1506	*1538*	224						1186	1457	988

Highest and lowest falls correct to 1990

November 1929

After persistently low rainfall during the first nine months of 1929, November was the wettest for sixty years and very little of England had less than 100 mm, while most of Wales, Devon and Cornwall had over 250 mm (10 in), including 340 mm (13·4 in) at Falmouth, 455 mm (17·9 in) at Tavistock and 747 mm (29·4 in) at Princetown on Dartmoor.

May 1932

Over England and Wales, the month was the wettest for 160 years and severe flooding occurred in the Don Valley at Bentley, South Yorkshire, and at Nottingham where part of the residential suburb of West Bridgford was severely flooded.

October 1939

This month cannot be ignored here on account of the very heavy rain which fell in Kent. By a sharp reversal of normal conditions, Kent was wetter than Snowdonia and the Lake District. A number of places in Kent exceeded 250 mm (10 in). Perhaps most remarkable of all was Margate's total of 261 mm (10·28 in), 25 mm more than fell in the whole of 1921. Our versatile climate can, therefore, in Kent at least, produce as much rain in a very wet month as may fall in the whole of a very dry year. Dover, in that same month, had 353 mm (13·9 in) and Folkestone 358 mm (14·1 in), amounts more suited to Borrowdale than to a drier part of England.

March 1947

It will long be remembered for its heavy snowfalls in the first half of the month and for heavy rains and the thaw which caused widespread flooding. Parts of the Fens were under water for weeks and seemed like an arm of the sea. The total at Kew of 118 mm (4·6 in) was the highest for the month since records began in 1866. Other high totals were 259 mm (10·2 in) at Falmouth and 283 mm (11·1 in) at Maiden Newton, Dorset. Further north, totals were lower with Stornoway actually having one-third of average.

January 1948

A very mild and stormy month with serious flooding in northern England. At Tynemouth the month's total of 170 mm (6·7 in) exceeded the previous record by 41 mm (1·6 in).

August 1952

A dismal summer month with widespread severe thunderstorms and as much as 229 mm (9 in) of rain falling in twenty-four hours over Exmoor, the cause of the Lynmouth flood disaster (see Chapter 6).

August 1956

Yet another August was also very wet. Rain fell on nineteen days at Kew and twenty-three at Keswick. Totals exceeded 125 mm (4·9 in) widely, with 172 mm (6·8 in) at Edinburgh, 271 mm (10·7 in) at Keswick and 276 mm (10·9 in) at West Kirby in the Wirral.

July to November 1960

In July rain fell every day over the Pennines. During this month, five depressions crossed the country and high rainfall totals were widespread over England and Wales, among them 303 mm (11·9 in) at Darwen, Lancashire, 180 mm (7·1 in) at Neath, West Glamorgan, 247 mm (9·7 in) at Malham Tarn and 235 mm (9·3 in) at Princetown, Dartmoor. August of the same year was very wet. Three depressions moved along the Channel and another three across the north Midlands. At Brighton and Kinloss, the rainfall was about three times the average. Serious flooding occurred near Exeter in October. Trains on the main line near Dawlish created a bow wave as they moved slowly along a track barely visible above the water. This month was remarkable for high rainfall totals at many places from Scotland to Devon. Craibstone, near Aberdeen, had rain on twenty-six days and a total of 237 mm (9·3 in). Over Dartmoor 380 mm (15 in) fell; at Exeter the total was 63 mm (2·5 in) less. An unusual feature of the month was the comparative dryness of the north-west Highlands with no water in some of the lochans on Rannoch Moor,[6] though as mentioned earlier good or bad weather is unlikely to occur simultaneously over all the British Isles, and this month proved no exception to that fairly reliable rule. As if these three very wet months, all occurring in one year, were not enough, November was also rainy with over 175 mm (6·9 in) at stations as far apart as Blackpool, Goudhurst in the Weald of Kent, Falmouth and Guernsey. Over Snowdonia the total was 624 mm (24·6 in), considerably above a year's supply of rain for Cambridgeshire, although Cape Wrath had only one-third of the average, and less than normal was registered over Northern Ireland and much of the north-west Highlands. But sunshine was above normal in many places.

November 1963 to September 1968

November 1963 was another reversal of the average, rather like October 1939. Many places along the south coast had over 200 mm (8 in), including 235 mm (9·3 in) at

Dover. (Apart from 1960, October 1967 was the wettest month over England and Wales as a whole since 1903.) Ramsgate achieved a Borrowdale-type total of 265 mm (10·4 in), and Eastbourne and Bournemouth had over 200 mm, and Penzance only a little less. At Eskdalemuir there was over twice the average and flooding was widespread. Rain fell on twenty-seven days at Ambleside where the total was 504 mm (19·8 in). Seathwaite's total was 775 mm (30·5 in); 1041 mm (41 in) was recorded at Fleetwith. July, August and September 1968 all had above average rainfall in England and there were the notable floods described in the next chapter.

Autumn 1976

The famous summer of 1976 came to an abrupt end at the end of August. September was very wet, probably the second wettest since 1727 over England and Wales, if we may rely on evidence provided by old records. Among high totals were: 193 mm (7·6 in) at Craibstone, near Aberdeen; 250 mm (9·8 in) at Chopwellwood, Durham; 198 mm (7·8 in) at Malvern; 217 mm (8·5 in) at Dover; 200 mm (7·9 in) at Blackpool; and 257 mm (10·1 in) at Hawarden Bridge near Chester. There were a number of heavy falls this month at Stokesley, North Yorkshire, and at Polperro, in Cornwall, where a flash-type flood following a thunderstorm surged through the town. Glasgow also had a record downpour of 84 mm (3·3 in) in three and a half hours on the 28th. On the 12th Durham had its wettest day on record when 87·8 mm (3·5 in) fell. Small regions of eastern England had less than 50 mm (2 in) and the northern Highlands and Shetland had less than average.

As usual, October proved the rule that no month seems to be very wet everywhere. Northern Scotland and Cumbria had less than average. For northern Scotland, that seems appropriate as the 1976 drought did not really affect that area. Most of South Wales, Snowdonia, Dartmoor, Exmoor, west Dorset and parts of the Isle of Wight all had over 200 mm. At Milford Haven there were 245 mm (9·6 in) and Dartmoor produced a total of 413 mm (16·3 in) at Princetown. Both November and December were also unsettled months, and the figures for Kew Observatory for the last four months of 1976 were September 107 mm (4·2 in); October 95 mm (3·7 in); November 83 mm (3·3 in); and December 67 mm (2·6 in); a total of 351 mm (13·8 in), or as much as had fallen in the previous twelve months.

October 1987

In the London area rainfall was three times the average and it was over twice the average across the Midlands and Wales, making it one of the wettest months of the century. From Cornwall across to Kent and over much of Wales more than 200 mm (8 in) of rain fell. Many houses in the village of Rottingdean, east of Brighton, were flooded by muddy water. Local farmers blamed the heavy rainfall for the floods, while others blamed the farmers for ploughing up land on slopes greater than 11°. There also appears to be a connection between the incidences of flooding and the change to cultivating winter wheat (sown in the autumn and wintering in the ground) in the 1970s. With this method of cultivation there is bare soil in the autumn until the seed germinates, so that heavy October rains can easily wash away the soil, which is what happened. A number of court cases followed; but the matter was settled out of court and so no definitive ruling on liability has been made. I make this point to remind readers that more floods is not necessarily an indication of climate change, but quite possibly the result of human activity.[7]

July 1988

A miserable summer month with only two-thirds of average sunshine and plenty of rain. Deep depressions often crossed the British Isles. The warmest day of the month was under 20°C (68°F) at Brighton and hardly anywhere exceeded 25°C (77°F) once. Rain fell on all thirty-one days at Helensburgh and over the Southern Uplands. There were twenty-eight days with rain at Sevenoaks, Kent, and twenty-four in parts of London.

Winter 1993–94

Between the beginning of September 1993 and the end of the following January, 658 mm (25.9 in) of rain fell near Chichester, which was way above average. The heavy rains at the end of December and in early January falling onto saturated ground caused flooding on the clay soils and also a substantial rise in the water table in the chalk Downs. The River Lavant rose to record heights and flooded part of Chichester and the A27 trunk road just to the east of the town. A temporary bridge had to be erected. Normally the Lavant is nothing more than a gentle trickle beside the A286 at Singleton and often dries up. In the centre and west of Chichester the river is in a culvert, but the volume of water from the heavy rains greatly exceeded the carrying capacity of the man-made channels. The river at Lavant has flooded before, though not in living memory.[8]

September 1995

Kinloss, in one of the drier parts of Scotland, received 295 mm (11.6 in) of rain during the month. There was serious flooding around the Moray Firth and Aberdeen. The railway line from Inverness to Aberdeen was unusable for some weeks because of flood damage. In the south of England rainfall was twice the average at Hurn (Bournemouth) and nearly twice at Heathrow. By contrast rainfall was half normal at Dublin. Pressure was low over the North Sea.

NOTES TO CHAPTER 5

1. Met.O. 617a, Part V, Asia, HMSO.
2. M.O., monthly averages, 1941–70.
3. M.O., monthly duration of rainfall in hours, 1951–60, and C.O.L. for 1985–94.
4. D.W.R. (Monthly Summary), and M.O., ten-year books of rainfall duration.
5. M.W.R. and B.R.
6. Observations by M.A. Town.
7. J. Boardman, 'Damage to property by run-off from agricultural land, South Downs, Southern England, 1976–93', *Geographical Journal*, vol. 161, 1995.
8. G. Holmes, 'The West Sussex Floods of December 1993 and January 1994', W. 50, pp. 2–7.

6 | Some Days Bring Deluges

Everyone is familiar with the persistent splash of a wet day, but few of us could say how much falls in these visitations. On a wet day recently the writer commented that 32 mm (1·26 in) had fallen, to receive the reply: 'I was driving through it and would have thought that there had been a foot of rain.' This remark is typical. To many people the mention of rainfall amount is meaningless. In Chapter 4 we established that an average rate of fall of about one millimetre an hour means steady rain, so a complete day of rain would be needed to achieve an inch (25 mm).

With annual rainfall we adopted 600 mm (about 24 in) per annum, London's fall, as a guide. In the same way, we considered 50 mm (2 in) a month as an average for the lowlands: 6 mm (0·24 in) is a reasonable fall in a day; 12 mm is not very common; and 25 mm occurs once or twice a year in the lowlands. Kew averages 1·3 days a year with rainfall exceeding 25 mm. At Manchester and Renfrew, the total is two, at Plymouth three and at Swansea four. Armagh is as likely to have 25 mm or more as is Oxford, the average being 1·6 days. At Kew the rainfall total on the wettest day of the year averages 32 mm (1·26 in); at Swansea, 42 mm (1·65 in). These facts give us our standard.

The official records of rainfall in Britain contain numerous instances of more rain falling in a single day than would be expected in a month. The rest of this chapter comments on some of the more sensational wet days of the twentieth century.

One of the most disastrous floods in this period occurred at the little town of Louth, Lincolnshire, on 29 May 1920.[1] The low hills of Lincolnshire had the full fury of the rain with 119 mm (4·7 in) in three hours at Elkington Hall, and 100 mm (3·9 in) at Horncastle. Louth itself recorded the comparatively modest total of 30 mm (1·2 in). The floodwater cut through the town drowning twenty-two people and causing much damage. It was thought that the floodwater had been dammed by debris and that the bursting of this obstruction had released the torrent.

A small but well-developed depression moved slowly northwards from the Straits of Dover to appear off Cromer in the evening of 25–26 August 1912. An intense rainstorm followed, giving more than 100 mm over 2700 km^2 (1000 sq. miles). Fig. 34 shows that the maximum fall occurred just to the east of Norwich,[2] where Brundall recorded 205 mm (8·1 in). Norwich itself experienced 185 mm (7·3 in) (or over a quar-

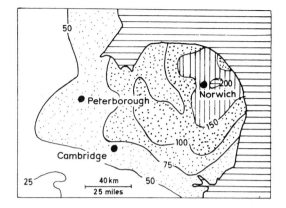

Fig. 34. **The Norfolk rainstorm of 25–26 August 1912** (rainfall totals in mm)

ter of the annual average) in one day. With such a deluge flooding was serious. On a wall in the city there is a tablet which records the high-water mark for the severe floods of the past. The highest of these floods bore the date 1614, but the vast flood of 1912 went 38 cm (15 in) higher. Much damage was done to roads and bridges throughout the county. A large area of land remained under water the following winter. It was fortunate that there were no steep slopes to accentuate the rush of water.

On 12 November 1897, Seathwaite had 204 mm (8 in) of rain. This constituted a daily record for some years, but on 11 October 1916 Kinlochquoich,[3] Inverness, registered 208 mm (8·2 in). That record only stood for a few months, for on 28 June 1917 a deluge caused by a small depression moving up the Channel made rainfall history by producing 243 mm (9·6 in) at Bruton, in the low-lying part of Somerset. A plaque by the parish church marks the flood level. In another part of the same county, at Cannington,[4] just west of Bridgwater, on 18 August 1924 a westerly wind bringing unstable polar air created a torrential downpour with 200 mm (7·9 in) falling in five hours, the final total being 239 mm (9·41 in).

It must not be thought strange that so many new records were broken in the early years of this century: records cannot be broken until there are records in existence to break. The normally dry east can expect a deluge of over 50 mm in twenty-four hours once in fifty years and 25 mm one year in two.[5] Over the mountains (e.g. Snowdonia), 150 mm in a single day can be expected twice a century; and 100 mm as frequently as one year in two.

In November 1929 there were severe floods in the Rhondda valley[6] on the 11th, when 211 mm (8·3 in) was measured at Lluest Reservoir. This is still the largest one-day fall measured in Wales, though it must be remembered that many mountain gauges are not read daily, so heavier falls may well have gone undetected, in the absence of an autographic rain gauge.

Dwellers in the Esk valley,[7] in North Yorkshire, had an unpleasant experience in July 1930. A slow-moving low was centred off Lincolnshire. Heavy rain associated with this depression fell on four successive days with the peak on the 22nd. On that day, 145 mm (5·7 in) fell at Castleton and 132 mm (5·2 in) at Danby. In four days Castleton had the remarkable fall of 304 mm (12 in), so that four months' normal precipitation came down in four days. An unprecedented feature was the use of the Whitby lifeboat at Ruswarp, 3 kilometres (2 miles) inland, where it was launched on to the floods to rescue trapped people.

The town of Boston, Lincolnshire, like all Fen localities, has a low rainfall of 595 mm (23·4 in) a year. Until 1931, the most sensational event in Boston's rainfall history was a fall of 78 mm (3·1 in) on 29 September 1883. On the morning of 8 August 1931 there was low pressure over south Norway and a slow-moving secondary depression over southern England, an ideal situation for heavy rain somewhere near the slow-moving system. Boston received 155 mm (6·1 in), or a quarter of a year's supply, of which 100 mm fell in about two hours. One would have thought that a long time would have elapsed before anything similar happened again. Yet only six years later, on 15 July 1937, 139 mm (5·5 in) of rain fell in twelve hours. Horncastle,[8] not far from Boston, had 184 mm (7·2 in) on 7 October 1960.

At 9 h on 15 August 1952, over 225 mm (8·9 in) of rain were recorded by the gauge at Longstone Barrow, Exmoor (Fig. 35).[9] A depression moving slowly east along the Channel (Fig. 36) brought heavy rain, which caused the flood disaster at Lynmouth. The month had already been wet and the peat and shales of Exmoor were quite unable to absorb the vast quantities of water, which subsequently surged down the valleys of the East and West Lyn rivers, carrying enormous boulders, washing away houses and hotels and sweeping 130 cars out to sea. Most of the flood surge was caused by the bursting of a log and vegetation dam, which released the pent-up waters down the short steep rivers. People living or holidaying in the area will never forget that night (14 August 1952) when darkness, accompanied by vivid flashes of lightning, came early because of the thick cloud. The power supply also failed and made the occasion even more alarming. Over thirty people died. A 7-ton boulder found in the basement of a hotel gives some idea of the force of the water (Fig. 37). A pile of boulders in the East Lyn river, probably deposited by a previous flood in 1769, was not moved in 1952, which suggests that the eighteenth-century storm was still more severe. Another flood was recorded in 1607. Exmoor's annual rainfall varies between 1250 and 2000 mm (49 and 79 in) a year. One-day falls of 200 mm or more are likely to occur at intervals of well over a century.[10] The intervals between the last three big floods were 183 and 162 years. But there is no guarantee that another flood may not occur in a shorter time, or that earlier floods were properly documented. Man may also have interfered with the streams, restricting the flow and thus increasing the risk of flooding. Caution is needed, therefore, in assessing changes in the frequency of flooding as indicators of climatic change (Fig. 174).

On 9 October 1967, Great Langdale, Cumbria, recorded 146 mm (5·7 in) of rain as the fronts of a small depression moved across the Lake District and south-west Scotland. A week later, 25 mm (1 in) of rain or more fell over two-thirds of England and Wales.[11]

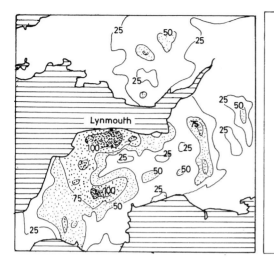

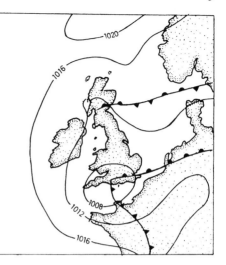

Fig. 35. **The Lynmouth Flood Disaster.** Rainfall totals (in mm) for the twenty-four hours to 9 h on 15 August 1952.

Fig. 36. **The synoptic chart for 18 h, 14 August 1952,** the evening before the Lynmouth Flood

Fig. 37. **A scene in Lynmouth after the flood**

Nine months after this widespread deluge, 37,000 km^2 (14,300 sq. miles) had over 50 mm (2 in) of rain on 11 July 1968. Many towns and villages in Devon were flooded or inconvenienced by the downpour. Rivers overflowed their banks and thousands of hectares of farmland were inundated, and as many as a thousand bridges were destroyed in a wide area from Devon to Lincolnshire. Fig. 39 shows where the heavy rain fell.[12]

People in other parts of the country were thankful to escape the worst of the weather, but the turn of East Anglia, the London area, Essex, Kent, Surrey and Sussex was to come on 14 and 15 September 1968, when a small depression deepened off the south-west approaches giving 50 mm of rain to the Scilly Isles. The low moved slowly east and the next day a pronounced frontal system became almost stationary over south-east England. 14 September was a day of heavy showers and some thunderstorms in Surrey, although they were not unusual. The morning of the 15th was very dark, with heavy rain and almost continuous peals of thunder. Such intense rain usually ceases soon, since the clearer conditions behind the front move from the west in the normal way. By 10 h many roads in Surrey and parts of Sussex were dangerously flooded and had obstructions, such as manhole covers forced up by the pressure of floodwater. Heavy rain continued throughout the day and by evening the floods were assuming disaster proportions. At Guildford, the recently built Yvonne

Fig. 38. **The scene in Millbrook, Guildford, on the afternoon of 16 September 1968.** Some shops at the bottom of the High Street were flooded to a depth of 2·4 metres (7·9 feet).

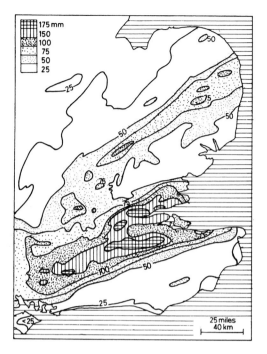

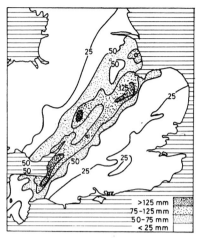

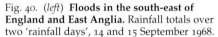

Fig. 39. (*above*) **Rainfall totals for the twenty-four hours to 9 h on 11 July 1968.** Widespread flooding occurred from Somerset across to Lincolnshire.

Fig. 40. (*left*) **Floods in the south-east of England and East Anglia.** Rainfall totals over two 'rainfall days', 14 and 15 September 1968.

Arnaud Theatre was flooded to a depth of a metre and was out of action for some weeks. A large new department store, built on the flood plain, had its basement filled up as if it were a water tank, the ground floor was inundated and much stock was lost. Damaged bargains were eagerly sought by shoppers during the following weeks. Fig. 38 shows the scene outside the store on the afternoon of the 16th. A bridge on the A281 Horsham–Guildford road was washed away at Shalford and the main Waterloo–Portsmouth line was interrupted by a collapsed bridge at Godalming. Severe flooding affected Molesey and Byfleet; but fortunately no rain fell over the headstreams of the Thames and Kennet so that a major disaster was avoided in London. Fig. 40 shows the distribution of heavy rain during this remarkable downpour. Most of the rain fell in the 'civil day', that is, between midnight and midnight, rather than in the 'rainfall day', which normally ends at 9 h and so the fall was not a record one.[13]

The year 1968 will long be remembered for 'the floods', and two major downpours in a year would seem to have been more than enough, but on 26–27 March the total rainfall exceeded 250 mm (9·8 in) over parts of the north-west Highlands of Scotland. The warm sector of a depression, containing very moist Tropical Maritime air, covered Scotland and it deposited much of its moisture over the mountains as the air was chilled to its dewpoint by forced ascent. Along the east coast from the Moray Firth southwards, the month's rainfall was about half the average. While the west of

Scotland was shrouded in low cloud, much of the south of England was enjoying magnificent weather with record March temperatures. Later in the year when southern England was very wet and pictures of floods appeared on the news, Scotland escaped.

The Highlands had another notable downpour on 17 January 1974 (Fig. 41). Again, much of the rain was orographic and Loch Sloy recorded 238 mm (9·4 in), which constitutes a record so far for Scotland, though there are two larger amounts for England. Between 25 and 27 December 1979, very heavy rain fell over the mountains of South Wales, with 140 mm (5·5 in) in the three rainfall days. On the 26th, Dowlais had 83 mm (3·3 in) of rain and Merthyr Tydfil 94 mm (3·7 in). The month had already been wet, and flooding was sudden and severe. In Cardiff many houses were inundated for the first time in living memory, with a metre or more of muddy floodwater.

Here are some record twenty-four-hour rainfall totals:[14]

279 mm	11·00 in	Martinstown, Dorset	18 July 1955
243	9·56	Bruton, Somerset	28 June 1917
239	9·40	Cannington, Bridgwater, Somerset	18 August 1924
238	9·37	Loch Sloy, Dunbarton	17 January 1974
229	9·00	Longstone Barrow, Exmoor	15 August 1952

Places in Dorset and Somerset hold the record for downpours in the history of British rainfall. One wonders whether these readings are genuine, and whether other larger totals have been omitted. What about Snowdonia, the Lakes and Scotland? Could these notoriously wet parts have beaten Dorset's cloudburst? Until we have a network of autographic rain gauges the largest twenty-four-hour total means the total in a rainfall day which normally ends at 9 h. In December 1954, Loch Quoich had 254

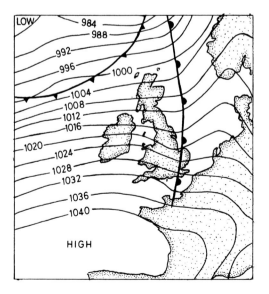

Fig. 41. **17 January 1974**. Record rainfall in the Highlands, in the warm sector of a depression. Loch Sloy recorded 238 mm (9·4 in). If the rainfall day had ended at a different time of day, the table above could have been different.

mm (10 in) of rain in 22·5 hours, but this was spread over two rainfall days. Otherwise Loch Quoich would have come second in the list. All the same, it is rather curious that the heaviest falls in one rain day have occurred in the south-west. Perhaps it is because these downpours are associated with thundery conditions, so that afternoon and evening storms are more likely to occur in one rain day there than in the Highlands, where rainfall is more likely to be associated with orographic uplift and fronts at any time of the day.

Details of two wet summer days, one of them at the height of the holiday season in August 1977, and the other of which spoilt an international sporting event in June 1978,[15] are shown in Figs. 43, 44 and 45.

Between 5 and 8 January 1982 over 100 mm (3·9 in) of rain fell in the three rainfall days over the Southern Uplands and the Pennines. Much of the rain fell on frozen and snow-covered moorland, so that there was no percolation. Severe floods followed, notably around York, where the river Ouse was 5 metres (16 feet) above normal. Ice floes became jammed under bridges and this caused further problems when the ice dam gave way. The effects of the flooding were made worse by the onset of a very severe spell of weather because the floods then froze over.

25 August 1986 was one of the worst bank holidays of the century. A depression deepened off south-west Ireland and moved north-east to the middle of the North Sea. Heavy driving rain and gales were widespread, though north Scotland escaped. Among high rainfall totals (in mm) were 25 (1 in) at Heathrow; 32 (1·3 in) at Boscombe Down on Salisbury Plain; 46 (1·8 in) at Elmdon, Birmingham; 56 (2·2 in) at Valley, Anglesey; 60

Fig. 42. **14 August 1996, Folkestone floods**. A total of 109 mm (4·3 in) of rain fell in Folkestone in twenty-four hours. 61 mm (2·4 in) of this fell in just seven hours, causing severe floods. People were rescued by boat from first-floor windows. A culverted stream added to the problems when it burst under pressure from the rainwater. The downpour was caused by a low-pressure system over the North Sea. Light winds over east Kent kept thunder-clouds overhead instead of driving them inland.

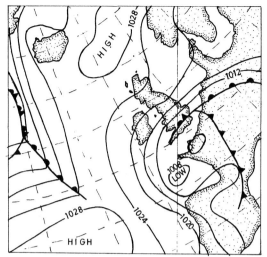

Fig. 44. (*above right*) **Chart for noon 17 August 1977,** one hour later than the satellite image in Fig. 43.

Fig. 45. (*facing page*) **10 h, 29 June 1978**. There were several low-pressure centres near Britain and there was rain throughout the daylight hours in the south-east. The large crowd that went to the Wimbledon Tennis Championships saw no play at all that day. In earlier years rain had washed out the final Saturday in 1972 and 1963. The chances of a totally wet afternoon preventing play are not very great: there have been 10 such days out of 627 since 1945 (Wimbledon Museum Library).

Fig. 43. (*above left*) **Satellite picture for 17 August 1977**. During 16 August a small but very active depression moved northwards towards Cornwall. Heavy rain lasted throughout most of the day with totals exceeding 25 mm (1 in) in south-western districts. A further depression affected south-east England on the 17th, giving cool weather with widespread heavy rain and thunderstorms. At Ruislip, west London, there were 113 mm (4·4 in) in 24 hours, with much flooding. The satellite image taken at 11 h on 17 August 1977 shows the cloud system of this depression. Most of Ireland and virtually all Scotland were not affected and had a pleasant sunny day.

(2·4 in) near Nottingham; and 68 (2·7 in) in Dublin. Grand prix racing in Birmingham, which was to have taken place on selected public roads, was cancelled. In Cheltenham the conditions were so wet that people got soaked running from pavement edge to restaurant. It was a very wet month altogether and ramblers walking the Cotswold Way found the going extremely muddy and slippery: it was like walking on chocolate semolina. One steep climb just east of Cheltenham could only be negotiated by holding onto a fence.

On 9 August 1993, as an anticyclone receded towards the Baltic, prolonged thun-

derstorms hit parts of Cornwall and Devon. Culdrose recorded 125 mm (4·9 in) of rain in nine hours to 9 h GMT, which caused floods in Helston.

NOTES TO CHAPTER 6

1. B.R. to 1920.
2. J. Glasspoole, B.R. 1912.
3. B.R. 1916.
4. B.R. 1924.
5. J.C. Rodda, 'A study of the magnitude, frequency and distribution of intense rainfall in the United Kingdom', B.R. 1967, pp. 204–15.
6. B.R. 1929.
7. B.R. 1930.
8. B.R. & M.W.R. 1931, 1937, 1960.
9. L.C.W. Bonacina, 'The Exmoor cataclysm', W. 17, p 334. A. Bleasdale and C.K.M. Douglas, 'Storm over Exmoor on 15 August 1952', M.M. 81, pp. 353–367. B.R. 1952.
10. M.D. Newson, *Flooding and Flood Hazard in the United Kingdom*, OUP, 1975, Ch. 1.
11. B.R. 1967.
12. P.R.S. Salter, M.M. 1968 and 1969.
13. B.R. 1968.
14. M.O.
15. All England Lawn Tennis Club.

7 | Thunderstorms

We wake to hear the storm come down,
  Sudden on roof and pane
The thunder's loud and the hasty wind
  Hurries the beating Rain.

Edward Shanks (1892–1953)[1]

The last chapter described unusually heavy and prolonged rains associated with fronts which were often slow-moving. This chapter will tell of very heavy rain from thunderstorms. Such falls differ from the other heavy downpours in that they are more localised, although the distinction is sometimes rather arbitrary. Often they are briefer, too. Certainly the heavy rain which affected Lynmouth was accompanied by thunder. But it was not a local storm like the one which broke over Hampstead on 14 August 1975 when 171 mm (6·7 in) of rain fell, although it was only over a few square kilometres that the storm was really severe.

Just what is thunder? It is the sound produced by the rapid and violent expansion of the air as it is heated by a lightning flash. The actual rumble of thunder depends on the time required for the sound to reach the observer's ear from various parts of the flash, which may be several kilometres long. In narrow valleys and densely built-up areas, echoes will sometimes increase the length of the rumble so that the thunder sounds almost continuous. It is rare for thunder to be heard more than about 20 kilometres (12 miles) away.[2] Lightning is really an enormous spark, with about 60% of the flashes occurring between clouds. For lightning to occur there must be an electrical charge generated. One hypothesis is that generation of charge within a cloud is connected with the fact that a water drop, when violently disrupted in a rapidly rising or falling air current, will split up, the largest fragments becoming positively charged and the fine spray negatively charged.[3] From this it seems that thunderstorms are only likely to develop in clouds which have reached a height of 16,000 to 20,000 feet (5 to 6 kilometres, or 3 to 4 miles)), where temperatures are in the ranges −20° to −40°C (−4° to −40°F). The anvil top which usually accompanies all thunderclouds, but cannot always be seen, is an indication that this has happened.

Sheet lightning is nothing more than the reflection of ordinary lightning. The popular distinction between the two is an illusion. All lightning is fork lightning. So-called 'summer lightning' is just distant flashes of fork lightning reflected by clouds. As light travels so fast, the flash is seen before the thunder is heard. Sound travels at about 1223 k.p.h. (760 m.p.h .) near the ground, which is about 12 miles a minute or 1 mile in five seconds. Hence, it is possible to obtain a rough estimate of the distance of the flash in miles, by dividing the time lag in seconds between flash and thunder by five. For the metric minded, dividing by three will give an approximation in kilometres.

If cumulus clouds develop rapidly from the middle of the morning and grow large, tall and dark, then showers are likely, often with thunder. If these clouds move quickly, the storms will probably only last a short time at any one place. More severe storms are possible if there are fewer, larger, slower-moving cumulonimbus clouds, which have sucked water vapour from a wide area. Often a distant bank of cumulonimbus presents an inspiring sight, but the people underneath are usually receiving a soaking.

For thunderstorms to occur, conditions must simply be those which favour the growth of very large cumulus or cumulonimbus clouds, that is, cumulus clouds giving rain, hail or snow. For the development of thunderstorms, a deep, moist, unstable layer of air is needed. The average distribution of thunder is shown in Fig. 46.[4]

Fig. 46. **Average number of days with thunder per year**

Thunder is generally associated in our minds with hot sultry weather. Certainly afternoon and early evening storms are prone to happen on the margins of a high-pressure system as a heat wave begins to collapse. All too frequently a thunderstorm 'changes the weather', or more accurately the change in the weather is responsible for the storm. The saying, 'three fine days and a thunderstorm' has a certain degree of truth in it, especially in unsettled summers.

A remarkably persistent pattern of fine nights, sunny mornings and afternoon thunderstorms, giving an equatorial style to the weather, became established in June 1970. Thunder was reported in England on twenty days. Even Scotland reported thunder on fourteen days. Many places escaped much rain. Yet it was a common experience in the Home Counties and Vale of Evesham to see awe-inspiring cumulonimbus clouds, complete with well-developed anvils, decorating the summer skies in an otherwise very sunny month. Some of these storms were very severe. Among notable falls of rain were 63 mm (2·5 in) in 117 minutes at Lossiemouth on the Moray Firth on the 7th; 62 mm (2·4 in) at Launceston, Cornwall, in 45 minutes on the 9th; and 83 mm (3·3 in) in two hours at Cambridge on the 17th. On the 11th, 45 mm (1·8 in) of rain fell in 20 minutes at Evesham. Later in the month 42 mm (1·7 in) of rain fell in only 10 minutes at Wisbech, a very rare fall. Heat storms of this type are most frequent in the warmer parts of the country. 'Heat storms' might be described as thunderstorms that occur in the afternoon and are not as a rule followed by a change in the weather: once the storm is over, the warm summer weather continues.

Most people have memories of being kept awake by night-time storms, such as Edward Shanks describes, though thunder is not very common during the night. However, these storms probably make a bigger impression because the thunder can be heard more easily since there are fewer sounds of human activity. Particularly frightening for some, night lightning brings eerie reflections round the bedroom. Sleep may well be difficult because of the high humidity and relatively high temperature. In the London area and over the south, as a rule, these storms are associated with low pressure over France or the Channel, and often last several hours. If the weather has been settled before the night storm, a change to cooler unsettled weather is likely. Usually these storms are frontal and are the result of convection currents brought about by cooler air undercutting warm, moist air.

I have heard many people comment quite incredulously on winter thunderstorms as if such things have never happened before. It is fair to say that the term thunderstorm here may be rather a misnomer, for the winter thunderstorm is frequently nothing more than a passing shower which gives only a single clap of thunder at any single point as it hurries on its way, usually south-eastwards across the country. The saying, 'Thunder in winter, Cold will bring', is liable to be apt since such storms will occur on a cold front where there is a marked contrast of temperature. Winter thunderstorms are commonest, however, in rather mild weather and they occur in air moving swiftly from the ocean.

The Diurnal and Seasonal Variation of Thunder [5]

Studies of thunder over a twenty-two-year period at Heathrow show that thunder is least likely between 2 h and 10 h in the morning, with about one occurrence in two years in each hour. It is most likely between 13 h and 18 h, during the heat of the afternoon. The secondary maximum between midnight and one o'clock in the morning is due to frontal storms from France. July is the most thundery month with thunder on three days, followed by August with two days. Taking 'thunderstorm hours', a phrase which means hours when thunder has been heard, July appears to be more than twice as thundery as August, which on four occasions out of the twenty-two had no thunder at all. Whichever way one measures thunder, July at Heathrow is more thundery and thunderstorms last longer than in other months. Some 90% of thunderstorms happen in the summer. With an average of fourteen days with thunder a year, only two are likely to occur outside the summer months.

At Ringway thunder is slightly commoner than at Heathrow. At first sight, it may seem rather surprising that London in the warmer part of the country does not have more thunderstorms. Using the word thunderstorm to mean more than just thunder in a quickly passing shower, London does indeed have more thunderstorms, but fewer days with thunder. August is definitely more prone to thunder in Manchester, the hours between 16 h and 17 h being the commonest time. Seasonal differences are summarised in Table 12.

Table 12 **Seasonal Variations in Thunder**

| | Mean Number of Hours | | | Mean Number of Days | | |
	Summer	Winter	Year	Summer	Winter	Year
Heathrow	29·1	2·9	31·9	13	2	15
Ringway	29·6	4·0	33·6	13	3	16

Some Notable Thunderstorms in Recent Years

The Martinstown Flood, 18 July 1955 [6]

Between 14·30 h on 18 July and 5 h on the 19th, over 150 mm (6 in) of rain fell in Dorchester and Weymouth. At Martinstown, 5 kilometres (3 miles) south-west of Dorchester, the total was 279 mm (11 in) (Fig. 47) which easily constitutes a record fall for a rainfall day. Even without this deluge, July 1955 would have earned itself a place in British meteorological history, as it was also very sunny and warm. Indeed some places had no rain at all (Chapter 10).

Throughout most of the month pressure was high to the north of Scotland. Easterly winds were blowing over England, while pressure was low over France. On the afternoon of 17 July, temperatures were in the range 28° to 29°C (82° to 84°F) in Dorset, while colder air was drawn in at high levels in the rear of a depression to the south and thunderstorms broke out over northern France and southern England. On 18 July, an active storm crossed the coast and slowed down, so that the rain which would normally have been spread over a wide area was deposited in one place.

The floods which followed in Weymouth, Dorchester and Bridport featured in the newspaper headlines. There were pictures of damaged cars and submerged caravans and boulder-strewn roads. But there was less loss of life than at Lynmouth: only two people were drowned. Though the rainfall was more concentrated than in the Lynmouth disaster, the porous nature of the chalk in the area enabled vast quantities of water to be absorbed. In fact, Martinstown suffered delayed-action floods, as the little river which borders the Martinstown to Winterborne Steepleton road rose rapidly the evening after the storm, and turned the road into a 4-metre-wide river.

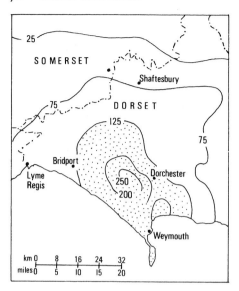

Fig. 47. **Rainfall (in mm) in the Dorset storm of 18 July 1955.** The twenty-four-hour total of 279 mm (11 in) at Martinstown, a few kilometres west of Dorchester, is a record for a one-day fall in Britain. In Dorchester the total was 183 mm (7·2 in).

The most severe structural damage directly connected with the storm was at Osmington Mills, east of Weymouth, where a small stream crosses the clay, and bridges and footpaths were washed away due to severe gullying.

As I have mentioned in the last chapter, the distinction between heavy rains from fronts and from thunderstorms is somewhat arbitrary. Although there was a depression not far away, the weather after the downpour soon returned to fine settled conditions.

Thunder, Great Heat and a Dust Fall[7]

1 July 1968 will be remembered by weather-men for its widespread dust fall; for its great heat, with many places in the south-east exceeding 32°C (90°F), with 34°C (93°F) at Liphook, Hampshire, in a light south-easterly breeze; and for its severe thunderstorms, often with large hail in the west and the Midlands, and parts of the north. Leeming, Yorkshire, had 35·7 mm (1·4 in) of rain in 8 minutes; at Slapton, Devon, there were 7 cm (2·8 in) hailstones. Large hail, several centimetres in diameter, also fell at Cardiff. Indeed, so many elements of the weather were involved to an unusual degree that it is very difficult to decide in which chapter the story of 1 July 1968 should be told.

At Guildford large orange blobs showed up conspicuously on parked cars, causing much annoyance to people who had recently cleaned them (Fig. 8). The dust fall covered an area from the Channel Islands to mid-Lancashire and north Yorkshire, though the mountains of Wales, the Cotswold Hills and most of the Sussex coast escaped. The orange blobs represented dust of Saharan origin brought down from high-level clouds by a few large raindrops. Places such as the Cotswolds, which were not covered by the overnight dust, had just sufficient rain to wash away the deposits brought down.

The dust fall was a most unusual phenomenon. Has such a thing happened before? Dust falls, we find, occurred in three consecutive years: 1901, 1902 and 1903. Some details are given here.[8]

On 22 February 1903, the temperature was high for the time of the year and the humidity was low, which was indicative of the desert, or Tropical Continental, origin of the air, in which winds had carried the dust to a great height. Much of the country south of a line from Anglesey to Wrexham and Ipswich was affected. At Findon, near Worthing, 'the colour of the dust was similar to brick dust'. At Edenbridge, Kent, 'there was a dry wind and the dust fall went on all morning, later turning to rain, when trees and shrubs showed signs of muddy rain and our windows were badly discoloured'. On the same morning, at Menheniot Vicarage at Liskeard, Cornwall, 'a thick orange-coloured fog came in off the sea. It was very different from our country white fogs'. About a year earlier, on 26 January 1902, 'the same red deposit came down in heavy rain during a funeral here and stained the open books and white surplices of the choir'.

Red-coloured rain fell over parts of west Scotland on 5–6 March 1977. The dust was probably of Saharan origin as high pressure over Europe was maintaining a southerly airstream over Britain. Another dust fall affected parts of northern England and the Republic of Ireland on 28–29 November 1979. The *Daily Mirror* described how gritty sandy rain speckled freshly cleaned motor cars in Blackpool. Over Cork and Dublin the dust was red. Pink dust like flour was visible on motor cars in parts of Northern Ireland and central Scotland on 29 June 1981. Have dust falls become more common, or are they more reported? The results of a dust fall show up so clearly on the rows of motor cars parked in our streets. There were other notable dust falls on 17 and 27 October 1984.

The Hampstead Storm, 14 August 1975[9]

On this day, 170·8 mm (6·7 in) of rain fell over Hampstead in about two and a half hours. Fig. 48 shows the distribution of rainfall. It was concentrated in a small area, the greater part of London remaining dry and even sunny. The afternoon weather was hot with temperatures as high as 29°C (84°F) and with quite a large amount of altocumulus castellanus (medium-level turreted cloud), later developing into large cumulus and cumulonimbus. Within half an hour of the storm breaking, the rivers Fleet, Westbourne, Tyburn and Brent, which are not normally noticed by the locals, flooded and huge waves pushed over garden walls and flooded homes, drowning a man in a basement. A parked car was washed away, turned on its side, and battered against a wall. Every house in Haverstock Hill was flooded, some to over 2 metres. Underground travellers were held up for two hours at Swiss Cottage because of water on the track and electrical failure, while above ground enormous traffic jams built up even well outside the area affected by the deluge, as pressure on the storm sewers forced up manhole covers and sent cascades of muddy water down the roads.

The London area lay between a slow-moving depression to the west and another over Germany, so that westerly and easterly winds met in the region of Hampstead where conditions were right for intense uplift, the 130 metres (427 feet) high hills probably adding extra uplift. The high temperatures and surface roughness would also have been contributory factors. There is some evidence to suggest that the height of Hampstead had something to do with this storm. It delayed the start of a Promenade Concert at The Royal Albert Hall as members of the orchestra were late. Since 1910 there have been seventeen occasions when over 40 mm (1·6 in) of rain has fallen in a rain day. The downpour of July 1923 when 65 mm (2·6 in) fell is the next highest total. At Kew during the same period more than 40 mm (1·6 in) of rain has fallen on twelve occasions.

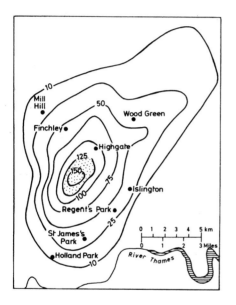

Fig. 48. **Rainfall (in mm) in the Hampstead Storm of 14 August 1975**

13 June 1979

During the afternoon a large, slow-moving cumulonimbus produced a severe local storm at Skipton-on-Swale, Yorkshire.[10] At Embsay Moor, 3 kilometres (2 miles) north-north-east of the town, 55 mm (2·2 in) of rain fell in fifty-three minutes. This is a rainfall intensity about fifty times the average lowland rate of fall. At Skipton Town Hall the total for the 13th was 35 mm (1·4 in), although most fell in one hour. Severe flooding followed. There were two feet of water in the General Post Office and four feet in some houses. Many parked cars were swept along by the torrent. One person was drowned, but a number were rescued, just in time. At Barden Reservoir, 6 kilometres (3·7 miles) to the north-north-east of Skipton, only 4·5 mm (0·18 in) of rain was recorded. Less than 1 mm of rain fell at Elslack, 5 kilometres to the west-south-west. Other severe floods occurred in the town on 1 July 1968 and on 3 June 1908.

June 1980

During the last week of the month complex deep low-pressure systems were centred near the north of Scotland and thunderstorms were widespread. During the afternoon of the 25th a particularly severe local storm broke over Sevenoaks, Kent. There were 116 mm (4·6 in) of rain in one and three-quarter hours, accompanied by an intense fall of hail which had to be cleared away by council workmen. Only 1·5 kilometres (1 mile) from the epicentre of the storm the rainfall total was 20 mm (0·8 in) and at Shoreham, 10 kilometres (6 miles) to the north, there was no precipitation at all.[11]

9 July 1981

A severe local thunderstorm gave 80 mm (3·1 in) of rain in seventy minutes to Littleover, Derby. A small area around the London Weather Centre had 58 mm (2·3 in) in fifty minutes.

5–6 August 1981

Heavy, thundery rain produced 95·9 mm (3·8 in) at Ringway Airport during the night, a record twenty-four-hour fall. The storms spread to the London area towards the end of the morning rush-hour. In three hours, 48 mm (1·9 in) of rain fell. At the height of the storm it was nearly dark and *The Times* next day had a photograph of the dome of St. Paul's silhouetted by street lights and by vivid lightning flashes at 10 h.

1 August 1991

Just after two o'clock on 1 August 1991 a localised severe thunderstorm affected part of Guildford. So heavy was the rain that manhole covers were forced up and there was local flooding, with water surging down the High Street and entering some shops. The author remembers the occasion well because he had purchased a small electronic rain gauge some weeks before and there had not been enough rain to test the accuracy of the instrument. At the height of the storm the rain was falling at the rate of 1 mm every 15 seconds, which would give a staggering hourly rate. A nearby standard gauge authenticated the total.

A moist south-westerly airstream covered much of England. Numerous small but very intense thunderstorms developed widely. Sheringham, Norfolk, recorded 45 mm (1·8 in) in an hour and a half. Tonbridge, Kent, and Pyrford near Woking had heavy falls of 22 mm (0·9 in) in half an hour.

NOTES TO CHAPTER 7

1. *Poets of our Time*, ed. Eric Gillett, Nelson, 1939.
2. R.J. Prichard, 'Audibility of thunder', J.M. 2, pp. 161–2.
3. Meteorological Office, 'Lightning', *A Course in Elementary Meteorology*, 2nd edition, HMSO, 1978, pp. 108–11.
4. H.H. Lamb, *The English Climate*, EUP, 1964, p. 73.
5. K.O. Mortimore, 'TORRO Thunderstorm Report', J.M. (monthly), C.O.L. (monthly). R.J. Prichard, 'Diurnal variations of thunder at Manchester Airport', W. 28, pp. 327–31.
6. A. Bleasdale, *The Dorset Storm*, M.O. Library, Bracknell.
7. C.M. Stevenson, 'The dust fall and severe storms of 1 July 1968', W. 24, pp. 126–32.
8. H.R. Mill & R.K.G. Lempfert, 'The great dust fall of February 1903 and its origins', Q.J. 30, pp. 57–73.
9. P. Beckman, 'Isohyetal map for north London, 14 August 1975'. J.M. 1, pp. 67–8. J.F. Keers and P. Westcott, 'The Hampstead Storm - 14 August 1975'. W. 31, pp. 2–10. R.A. Tyssen-Gee, W. 31, p. 278.
10. *Craven Herald and Pioneer, The Telegraph and Argus (Bradford)*, 'Deluge near Skipton, Yorks'; also J.M. 5, pp. 10–15.
11. P. Rogers, 'Sevenoaks Storm of 25 June 1980', J.M. 5, p. 223.

8 | Hail and Hailstorms

In the shower cloud, snowflakes impinge on supercooled water droplets (water with a temperature below 0°C), causing them to freeze immediately and turn into soft ice pellets or hailstones.[1] Strong up-draughts and down-draughts in cumulonimbus clouds are ideal conditions for the formation of hailstones (Fig. 49). Showers are more frequent over windward coasts and hills in Maritime Polar and Arctic airstreams, so it is not surprising that these areas have more days when hail is reported. Over the ten years 1966–75 [2] the average number of days with hail was eighteen at Benbecula in the Outer Hebrides (fully exposed to instability showers) but only nine at Stornoway on the eastern side of the larger island of Lewis. At Edinburgh, sheltered from both airstreams, hail is infrequent, falling on only two days a year. At Kew the total is three. The observatory at Edgbaston reports eleven days with hail, for it is more exposed to shower activity as it sits on a significant range of hills exposed to north-west winds carrying showers through the Cheshire Gap. In the more sheltered parts of Northern Ireland hail is probable on ten days at

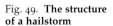
Fig. 49. **The structure of a hailstorm**

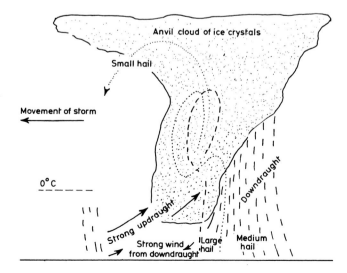

Fig. 50. **A deep fall of hail at Tunbridge Wells, 6 August 1956**. The town narrowly missed two more severe hailstorms in May 1922 and June 1908. The 1956 storm rated H4 on the Hail Intensity Scale. The 1922 storm rated H6 and the 1908 H5.

Armagh. In the south-west of England, Falmouth is likely to have hail on three days.

At Newquay the total is ten in an average year. Curiously enough, Jersey Airport records hail on twenty days a year, which seems rather surprising, for the climate of the Channel Islands is among the most favoured in the British Isles and one does not hear of hail being rated as a hazard to the glasshouses or market-garden crops grown there. The explanation is simple. Hail is associated with showers, and north-west winds must cross the relatively warm waters of the English Channel to reach Jersey, so that such air will tend to be unstable, producing quickly passing showers with only small hail. Nor must it be forgotten that a day with hail may mean a day on which the weather-station observer happened to spot a few small hailstones. The number of days with recorded hail is likely to be greater at weather stations manned round the clock, i.e. synoptic stations.

Over the hills and coasts of the west and north, hail has been shown to be comparatively frequent, but the hailstones are usually small. Over the lowlands of the east and south hail is infrequent, but the hailstones are often larger and associated with slower-moving summer thunderstorms.

Fig. 50 shows the aftermath of such a storm: a bus is pushing its way through deep drifts of hailstones in Tunbridge Wells after the hailstorm of 6 August 1956. It is probably events like this which give rise to the legends of summer snow in lowland Britain, for at first sight it does look like snow. There, however, the similarity ends since hail can occur in quite high temperatures, of 15°C (59°F) or even more, much too high for any snowflake to survive a fall through the air, let alone lie on the ground. Large cumulonimbus clouds were visible from Tunbridge Wells about ten in the morning, very early in the day for clouds of this type. An hour later heavy rain had begun, with large hail about 11.30 h, which accumulated in great drifts. Thunderstorms were widespread over many parts of the south-east, and fields of cabbage were reduced to shreds. There were several reports of people out for the day at the coast who, after experiencing plenty of sunshine, returned home to find their gardens ruined or under water.[3]

Fig. 51. **5 September 1958. Giant hailstones at Horsham, West Sussex.** Tennis ball on right.

Some of the largest hailstones[5] recorded in Britain fell near Horsham, West Sussex, on 5 September 1958 (Fig. 51). Giant hailstones also begin their lives as snow- or ice-crystals; they increase rapidly in size by collision with supercooled droplets during their journeys up and down in the violent air currents within the storm cloud, eventually becoming too heavy and falling out of the cloud to the ground. An eyewitness from the village of Kirdford, 16 kilometres (10 miles) south-west of Horsham, says: 'At Hillsgreen Farm[6] 10,000 bushels of apples from 50 acres were destroyed, the remainder left on the trees were nearly all hail-marked and worthless. The day had been fine and sunny, but about 16.45 h GMT a black mass of cloud approached from the west, and by 17 h a storm of wind, hail, thunder and lightning was raging, which lasted half an hour. It was difficult to see what was happening as the air was full of flying leaves, twigs, branches, hail and rain. This house, though on a hill, was soon invaded by water as every drain was blocked by debris or hail. I was able to get out about 18 h and measured several hailstones averaging 18 mm (0·7 in) across. The garden was a scene of desolation, though the village a kilometre away escaped the worst. There were also reports of small pits left in lawns where hailstones had melted.'

On 9 July 1959, large hail fell at Wokingham, Berkshire. This severe thunderstorm, like so many, travelled against the surface wind. For the wind at low levels was from the north-east, but at high levels from the south-west, so that there was a strong flow of air near the ground towards the storm in front and at high levels in the rear, relative to its movement. As the storm approached, the sky became very dark and, at the

Table 13 **The TORRO Hailstorm Intensity Scale**[4]

H-scale number	Effects	Range of size codes
0	True hail of pea size, but no damage.	1
1	Leaves holed/flowers damaged.	1–3
2	Leaves stripped from trees and plants; vegetables shredded; fruit and crops bruised and scarred.	1–4
3	Some panes in glasshouses/skylights broken; wood fences scored. Caravan bodywork dented. Canvas and perspex holed. Stems of crops severed and seeds threshed.	2–5
4	Some house windows, vehicle windscreens broken; glasshouses shattered; some felt roofs pierced; paint scraped off walls. Small branches broken from trees. Ground pitted. Unprotected birds and poultry killed.	3–6
5	Some roof tiles/slates broken; many windows smashed; plate-glass roofs broken; brick walls pitted. Car bodywork visibly dented. Strips of bark torn from trees. Large branches broken down. Risk of serious or fatal injury to small animals.	4–7
6	Plate/reinforced glass windows broken; many roof tiles/slates broken; concrete walls/pavements pitted; corrugated iron and some other metal roofs holed. Light aircraft damaged.	5–8
7	Roofs severely cut up; window frames carried away. Bodywork of cars and light aircraft severely damaged. Ground deeply indented and scoured.	6–9
8	Bodywork of cars and light aircraft destroyed; severe damage to commercial aircraft. Small tree trunks split apart. Risk of serious injury to persons caught in the open.	7–10

same time as a low banner of cloud arrived overhead, the surface wind suddenly changed to become strong south-westerly, followed by a fall of large hail with very little rain for about five minutes, and then by small hail and rain.[6]

On 13 July 1967 a shallow depression over northern France was moving slowly northwards. There appears to have been a connection between shallow lows to the south and damaging hail. Melksham, Chippenham and Bradford-on-Avon, Wiltshire, were badly affected. There were reports of 2 oz hailstones (57 g) which smashed windscreens, dented car bodywork and ruined glasshouses.

The temperature reached 33°C (91°F) in London on 1 July 1968 and a small depression moved along the front separating very hot continental air from much cooler maritime air. At Pontypridd, South Wales, holes of 50 mm (2 in) diameter were found in roofs and many car and house windows were broken, giving the appearance of big bullet strikes. The hail swath stretched from east Devon (70 mm (2·8 in) hailstones at Slapton) to South Wales. A similar tale of damage occurred the next day in Shropshire as the cooler air spread north and east.

A severe hailstorm affected Ludlow, Shropshire, on 22 June 1982. A photograph in my possession of a man shovelling hail from waist-high drifts merits inclusion; but on balance the Tunbridge Wells hailstorm photo is more dramatic. (The hailstorm also occurred in 1956. Too many recent illustrations of unusual events might be seen to be indications of climatic change.) Part of Bristol had knee-deep hail on 16 June 1982.

On 5 June 1983 an almost stationary front on the French side of the Channel separated very hot air over France from somewhat cooler air coming round the flank of an anticyclone over Scotland. Hailstones left the vinyl roof of a caravan at Church Knowle leaking like a sieve. In Poole, Dorset, many cars were dented. Roofing tiles were broken at Lee-on-the-Solent, Southsea. Hayling Island and Chichester had similar damage. Over much of the country it was a glorious summer day. The next day hailstones pierced roof gutters along the entire length of one road in Manchester.

22 August 1987

Over south-east England temperatures approached 30°C (86°F) in the afternoon. Pressure was high to the east and a small low developed over England. The onset of the storm in Chelmsford was heralded by a violent squall which tore large branches from trees. 41 mm (1·6 in) of rain was recorded in 30 minutes and the eastern and northern districts of the town were subjected to intense hail of 6 to 10 mm (0·2 to 0·4 in) diameter, which caused considerable damage to gardens. At West Bergholt hail the size of walnuts fell almost continuously for half an hour from 15.15 h, covering the ground for up to three hours. In the Stour Valley one motorist, observing the windscreen bulge with the impact of the hailstones, described the storm as resembling 100 hammers hitting the car simultaneously. At Dedham hail lay 15 cm (6 in) deep, reminding older residents of a similar storm in June 1942.

Glass roofs and skylights experienced widespread breakages, causing flooding. St. Audry's Hospital at Melton sustained very severe damage: every skylight was at least partially damaged, while an average of twenty panes of glass were broken in each of the wards, many of which suffered flooding.

There was much damage to vehicles. Bodywork was pitted, some cars with hundreds of dents. The widespread damage to vehicles may reflect the uncanny way in which the storm seemed to follow the very busy A12. At Ardleigh 80 acres of orchards were destroyed. In Woodbridge the contents of most gardens were flattened.

According to TORRO there have been six hailstorms of similar magnitude this century.[7] The path of the 1987 storm parallel to the coast suggests that a sea breeze convergence zone may have contributed to the development and path of the storm.

18 September 1992[8]

Hundreds of dead sea-birds, chiefly black-headed gulls, were found floating in the sea 10 kilometres (6 miles) off the Essex coast. At the request of the Royal Society for the Protection of Birds, some of the dead birds were sent to the Veterinary Investigation Centre in Cambridge. Their death seemed a complete mystery, as no evidence of pollution was found. One theory is that the birds could have been killed by the very violent hailstorm which hit the area. Foulness Island bore the brunt of the storm in the early hours.[9] The Shoeburyness Meteorological Office recorded an increase in windspeed from calm to 34 knots in less than a minute. On the 17th the temperature was in the low 20s °C during the day but much cooler air spread in overnight.

Other Notable Falls of Hail

More details of giant hail falls are given below. Further research would produce more hailstorms worthy of note, though the area affected by large hail 25 mm (1 in) or giant hail 37 mm (1·5 in) diameter is usually very small. The vast majority of people in Britain will probably never see damaging hail. David Atkins, author of *Tales of a Sussex Orchard*,[10] was not so lucky. His fruit was severely damaged twice: in 1958 and on 31 May 1964. The farm lies a few kilometres north of Pulborough, West Sussex.

Damaging hail is most likely to occur in a belt from Sussex across to Northamptonshire.[11] Just how rare is damaging hail at any one place? A village in the south-east Midlands could anticipate six such storms within a radius of 16 kilometres (10 miles), per century, which does not sound quite so rare: it is once every sixteen years.

- *Nottingham*, 3 September 1916, 32 mm (1·3 in) hailstones.
- Hailstones weighing 450 g (16 oz) were reported to have fallen at *Plumstead*, London, in July 1925.
- Many panes of glass were broken at *Peebles* in August 1930 when hailstones of 37 mm (1·5 in) diameter fell. Holes made in corrugated iron left evidence of the size.

- On 12 August 1938, hail lay deep on the ground at *Wold Newton*, near Bridlington, Yorkshire, remaining until the next morning.
- *Leyland*, Lancashire, had 31 mm (1·2 in) hail on 30 June 1944.
- 2 July 1946. Hailstones between 20 and 30 mm (0·8 and 1·2 in) did much damage to crops and glass in *East Anglia*.
- 3 July 1946. There was much damage to crops and glass in the *Farnham* area of Surrey.
- There was large hail at *Camborne*, Cornwall, and *Hurn*, near Bournemouth, on 2 November 1962.
- Hail did much damage to farm crops and market gardens in *Cambridgeshire* on 27 June 1970.
- In *Shetland* there was a severe thunderstorm accompanied by large hail on 16 August 1973, small hail being common there but large hail distinctly rare.
- 14 July 1975. Many cars were dented in *Sutton Coldfield*, and greenhouses were wrecked and verandah roofs shattered during an early morning storm.
- In south *Devon* skylights and cars were damaged by hail on 13 December 1978.
- 26 May 1985. In *Essex* and *Suffolk* there was a hail swath 70 kilometres (43 miles) long and 15 kilometres (9 miles) wide, where hailstones of 10 mm (0·4 in) or more fell. Widespread damage occurred on west-facing windows and roofs. There was a gentle south-easterly flow, veering south-west.

NOTES TO CHAPTER 8

1. F. Ludlam, 'The hailstorm', W. 16, pp. 152–62.
2. M.W.R. for all hail data (unless otherwise listed). D.L. Champion, 'The seasonal distribution of hail in Great Britain', W. 3, pp. 201–5.
3. R.E. Booth, 'The Tunbridge Wells Hailstorm', M.M. 85, pp. 297–9.
4. J.D.C. Webb, D.M. Elsom, G.T. Meaden, 'The TORRO Hailstorm Intensity scale', J.M. 11, pp. 337–9.
5. I. Holford, *Guinness Book of Weather Facts and Feats*, Guinness Superlatives, 1982, p. 189.
6. E.H. Rowsell, 'The storms of 5 September 1958'. B.R. III, 1958, pp. 15–21.
7. J.C. Webb, 'The hailstorms and thunderstorms of 22 August 1987 in Essex and Suffolk', J.M. 15, pp. 305–14.
8. 'Birds and animals die in British hailstorm, 18 September 1992', J.M. 18, p. 62.
9. M.W.R. and J.M. J.D.C. Webb, 'Britain's severest hailstorms and "hailstorm outbreaks", 1893–1992', J.M. 18, pp. 313–27.
10. David Atkins, *The Cuckoo in June: Tales of a Sussex Orchard*, The Toat Press, Pulborough RH20 1DA, 1992.
11. J.D.C. Webb, M.W. Rowe and D.M. Elsom, 'The frequency and spatial features of severely damaging British hailstorms', J.M. 19, pp. 335–45.

9 | Twisters

Year	Number of tornadoes reported
1001–1200	6
1201–1400	12
1401–1600	9
1601–1700	18
1701–1800	56
1801–1900	214
1901–1975	724
1976–1995	590
Total	1629

Table 14 **Tornadoes**

The twister, a dark and sometimes sinuous—often cone-shaped—trunk hanging between cloud and ground, moving fairly fast, is a familiar sight in the Middle West of the United States, and many excellent photographs exist of these destructive tornadoes. The American word 'twister' is a very apt translation of 'tornado', because the air rotates or twists very rapidly within the funnel, creating very low pressure in the centre. Some people confuse tornadoes with tropical storms, the West Indies hurricane, the cyclone of the Indian Ocean and the typhoon of the Pacific. Tornadoes have two things in common with tropical storms: low atmospheric pressure and strong winds. By contrast with tropical storms, the tornado is a minute affair with a diameter sometimes less than 100 metres, compared to hundreds of kilometres for the others. So low is the air pressure within the funnel that buildings sometimes explode or collapse outwards, as the air pressure outside the building drops extremely rapidly while that inside remains momentarily higher.

Many American houses in the Middle West have special tornado shelters where people can wait until the furious wind has passed by and it is safe for them to come up. Luckily, tornadoes are not very common and normally not severe in Britain. Our tornadoes are still so unusual that an observant person, even a keen meteorologist who would like to do so, may never see one. Compared to those in North America, shown so well in the opening scene of the film *The Wizard of Oz*, ours are relatively gentle, damaging roofs and felling trees, rather than flattening houses. They are mostly short-lived, and unlikely to last more than a minute or two at any place unfortunate enough to be affected. The destructive path may be anything between a few hundred metres and 50 kilometres (30 miles) or more in length.

Fig. 52.
**The Bedfordshire
tornado of
8 June 1955**

If a tornado moves or develops over water, then it is called a waterspout. Fig. 52 shows a tornado near Bedford. Possibly this tornado would be the equivalent of Force 2 or 3 on the TORRO Tornado Intensity Scale, shown in Table 15.[1]

The figures in Table 14 quote known tornado occurrences in Britain going back to the eleventh century.[2] These figures will produce all kinds of reaction, such as: 'Yes, I knew it, aeroplanes and space contraptions are affecting our weather.' There were more tornadoes recorded in the first seventy-five years of this century than in the previous 800 years. But of course people did not report all the tornadoes that occurred in the olden days. Records would not have been kept and there would probably have been no one interested in collating the sightings of such phenomena. Obviously the eighteen tornadoes reported in the seventeenth century must have been particularly severe. Many others must have passed totally unnoticed, like those in Norfolk in December 1975, which came to light only by chance, as mentioned later (page 93). So it is too early to say whether tornadoes are increasing in frequency as some suspect, particularly over built-up areas where strong warm air currents can trigger off a vortex.[3]

Table 15 The TORRO Tornado Intensity Scale

TORRO Force	Name	Characteristic damage on the tornado scale
FC	FUNNEL CLOUD OR INCIPIENT TORNADO	No damage to structures, unless to tops of tallest towers, or radiosondes, balloons, aircraft. No damage in the country, except possibly to highest treetops and effect on birds and smoke. Record FC when tornado's spout seen aloft but not known to have reached the ground.
		Subdivisions A, B, C apply to urban situations, D to rural situations.
0	LIGHT TORNADO	A Loose light litter lifted from ground in spirals.
		B Temporary structures like marquees seriously affected.
		C Slight dislodging of least secure and most exposed tiles, slates, chimney pots and TV aerials may occur.
		D Trees severely disturbed, twigs snapped off. Bushes may be damaged. Hay, straw and some growing plants raised in spirals.
1	MILD TORNADO	A Heavier matter levitated include planks, corrugated iron, deck chairs, light garden furniture, etc.
		B Minor to major damage to sheds, outhouses, locksheds and other wooden structures (such as henhouses).
		C Some dislodging of tiles, slates and chimney pots.
		D Hayricks seriously disarranged, shrubs and trees may be uprooted, damage to hedgerows, crops, trees.
2	MODERATE TORNADO	A Exposed, heavy mobile homes displaced; light caravans damaged.
		B Minor to major damage to sheds, outhouses, lock-up garages, etc.
		C Considerable damage to slates, tiles and chimney-stacks.
		D General damage to trees, big branches torn off, some trees uprooted. Tornado track easily followed by damage to hedgerows, crops, etc.
3	STRONG TORNADO	A Mobile homes displaced, damaged or overturned; caravans badly damaged.
		B Sheds, lock-up garages, outbuildings torn from supports/ foundations.
		C Severe roof damage to houses, exposing much of roof timbers, thatched roofs stripped. Some serious window and door damage.
		D Considerable damage (including twisted tops) to strong trees. A few strong trees uprooted or snapped.
4	SEVERE TORNADO	A Caravans and mobile homes destroyed or gravely damaged.
		C Entire roofs torn off some frame/wooden houses and small/medium brick or stone houses and light industrial buildings, leaving strong upright walls.
		D Large well-rooted trees uprooted, snapped, or twisted apart. Soft ground possibly furrowed by a tornado spout to a depth of about a metre.
5	INTENSE TORNADO	C More extensive failure of roofs than for force 4, yet with house walls remaining. Small weak buildings, as in some rural areas (or as existed in medieval towns), may collapse.
		D Trees carried through the air.

Table 15 **The TORRO Tornado Intensity Scale** *continued*

TORRO Force	Name	Characteristic damage on the tornado scale
		Subdivisions A, B, C apply to urban situations, D to rural situations.
6	MDERATELY DEVASTATING TORNADO	A Motor vehicles over one tonne lifted well off the ground. C Most residences lose roofs and some a wall or two; also some heavier roofs torn off (public and industrial buildings, churches). More of the less strong buildings collapse, some totally ruined. D Across the breadth of the tornado track, every tree in mature woodland or forest uprooted, snapped, twisted, or debranched.
7	STRONGLY DEVASTATING TORNADO	C Walls of frame/wooden houses and buildings torn away; some walls of stone or brick houses and buildings collapsed or are partly beaten down. Steel framed industrial buildings buckled. Locomotives and trains thrown over.
8	SEVERELY DEVASTATING TORNADO	C Entire frame houses levelled; most other houses collapse in part or whole. Some steel structures quite badly damaged. Motor cars hurled great distances.

Fortunately tornadoes of force 9 and 10 have not occurred in Britain.

In Britain, tornadoes seem most likely to follow a north-easterly course, generally running parallel to an advancing cold front. They may occur in any season and are most common in the afternoon in cool showery conditions. (Or are they more readily observed then?) Deep instability, that is, a broad layer of air with a desire to rise, a moist layer of air near the ground and much colder air above, and a cold front nearby, seem basic requirements for tornado development.[4] Some reports of these unusual disturbances are given here. It is not claimed that the list is anything like comprehensive in its coverage.

Some Notable Tornadoes

Violent storms in association with a cold front produced a tornado of force 6 on the TORRO Scale in South Wales on 27 October 1913. There was a narrow band of destruction some 18 kilometres (11 miles) long.[5] Birmingham experienced a severe thunderstorm on 14 June 1931 at about 14.30 h, and about the same time a tornado, also of force 6, developed in Sparkhill, passing through Greet, Small Heath and Bordesley to Erdington. Much damage was done to houses and many trees were uprooted over a distance of 19 kilometres (12 miles). Another tornado was reported in the Birmingham area on 4 February 1946.

On 7 July 1938 a tornado formed about 3 kilometres (2 miles) north of Boxmoor in the Chiltern Hills. The track was about 90 metres (300 feet) wide and extended for 15 kilometres (9 miles). Near Boxmoor a stationary sidecar and motorcycle were blown across the road. The path of the tornado resembled a battlefield, with haystacks being torn apart and many trees felled. There were instances of adjacent trees lying in opposite directions, an indication that the funnel passed between them, as the wind blows in opposite directions on either side of the funnel.[6]

21 May 1950 was a particularly stormy day and a tornado, force 6, associated with large hailstones and thunderstorms, developed near Wendover, Buckinghamshire, and moved north-east to Norfolk.[7] 'It felled five large trees in a field just to the south of the town, damaged roofs and raised a column of water from a nearby canal. At Aston Clinton, 4 kilometres (2·5 miles) to the north-east, the funnel cloud was seen to be following a zig-zag track, possibly being steered by minor topographical features, and severe damage was done to a farm and school buildings in the village. The funnel cloud now lifted well clear of the ground, but lowered again about 8 kilometres (5 miles) further on at Puttenham, where a brick barn was demolished, and where steel girders supporting the corrugated iron roof of another barn were bent and twisted, part of the roof being found in the top of a tree nearly a hundred metres away.' The built-up areas worst affected were Linslade, Buckinghamshire, and Leighton Buzzard, Bedfordshire. Just to the south of Linslade, the path of the tornado was about 45 metres (150 feet) wide and many large trees were felled; in the town some fifty roofs were badly damaged (one being completely removed), and a bakery was demolished. Some parked cars were lifted and carried several metres. A horse box with a horse in it was also lifted.

On 19 May 1952, a tornado moved from north to south from Hardstoft to Tibshelf, near Chesterfield, Derbyshire, damaging thirty-two houses.[8] It was observed soon after 14.30 h. The sky just before the event was dark, with a shower of large raindrops and a few big hailstones up to 16 mm (0·6 in) across. An eyewitness tells of the top of a wooden building, a few metres away, suddenly bursting upwards, the loose wood being sucked out. As with many tornadoes a cold front was near, though the possibility of warm moist air from cooling towers and blast furnaces, where the track began, triggering off the vortex cannot be entirely discounted.

Widespread winter thunderstorms, not consisting merely of a passing clap of thunder, and a destructive tornado occurred on 8 December 1954.[9] First observations of a funnel cloud came from Charmouth, Dorset, and from the Isle of Wight between 15.30 and 16 h. Damage was reported from Havant (north-east of Portsmouth) where boats were hurled across a road. At Chiddingfold, Surrey, a 24-metre (78-foot) pine-tree snapped and chicken houses were lifted into the air. The line of damage was not continuous as the funnel probably did not reach the ground again until Bushey Park, near Hampton Court, where trees were blown to the ground. Between 16.30 and 17 h the tornado swept through Chiswick, Gunnersbury, Acton and Willesden leaving many damaged houses and shops in a zone some 16 kilometres (10 miles) long and 365 metres (1200 feet) wide. Associated with the tornado was a severe hailstorm and very vivid lightning. At Acton, a car was sucked 5 metres into the air, the wall of a large Victorian house fell outwards, and a factory was virtually destroyed. More tornadoes occurred in the London area on 29 November 1970 at Barnet, and on 25 January 1971 at Avery Hill, Eltham, Welling and Upminster. The latter storm seems to have been of force 6 on the TORRO Scale.

The gardens of the Royal Horticultural Society at Wisley near Woking in Surrey, were visited by a tornado at 15.45 h on 21 July 1965.[10] The track was about 5 kilometres (3 miles) long, from Pyrford south-eastwards across part of the fruit trial grounds. Remarkable evidence of the very low pressure to be found in the middle of the vortex was afforded by an apple tree which was sucked out of the ground, its root system virtually intact, and with minimal damage being done to the tree. Slow-moving depressions were situated near the British Isles and there were widespread thundery showers.

The north-eastern suburbs of Coventry had a tornado at about 16 h on 21 April 1968. An hour before the storm it was calm and sunny. Immediately before it there was a short thunderstorm with hailstones up to 25 mm (1 in) in diameter. The tornado began at Wyken Grove and continued north-eastwards, missing Barnacle village by a few hundred metres. A number of cars and caravans were damaged or destroyed. Afterwards the weather was much cooler and fresher.[11]

On 26 September 1971 people in Rotherham, South Yorkshire, reported seeing a cloud shaped like an ice-cream cornet at about 16 h. At Rawmarsh much damage was done to the roofs of some houses on an estate. There was a report of an Alsatian dog in its kennel being carried over a fence and of garden railings being sucked out of the ground. Evidence provided by damage suggests that the tornado's track was about 20 kilometres (12·5 miles) long.[12]

Around eleven at night on 1 December 1975, seven tornadoes occurred in East Anglia. Though considerable damage was done there were no accounts of them in the local or national press. Five days later, one of the people affected by the damage reported the matter to the Anglia Television Weather Office. As a result, an investigation was made and evidence of seven distinct tornado tracks was found. If the nature of the damage had not been examined, it is quite possible that the whirlwinds of that stormy winter's night would have gone unrecorded. Among the more spectacular events were turnips being sucked out of the ground and a piece of iron roof-sheeting over 2 metres (6·5 feet) long being hurled javelin-like against the door of a farmhouse, terrifying a babysitter. One person describes how he woke up next morning to find only a concrete pad and an oil drum where his garage had once stood.[13]

During the afternoon of 8 October 1977, a tornado (TORRO force 3) travelled north-north-west just to the east of Grantham and Newark.[14] Only little more than a kilometre of its destructive track lay across a built-up area, where about thirty houses were seriously damaged. In Grantham a complete garden shed was carried over the tops of houses and dumped on a main road. At Stubton a large shed filled with hay, two more sheds and two hen-houses (with hens) were lost. Generally the line of damage was not more than 15 metres (50 feet) wide. This twister was reported as making a noise 'like very heavy lorries coming up the garden path'.

Newmarket was affected by a tornado (TORRO force 4 or 5) during the early morning of 3 January 1978. There was damage in a 200-metre strip extending east towards Bury St. Edmunds. A number of houses lost nearly all their tiles. As many as ten other tornadoes occurred on the same night, in association with a southward-moving cold front.[15]

The village of Llandissilio, Dyfed (Pembrokeshire), had a tornado (TORRO force 4) on 12 December 1978 during the early morning. The funnel sucked off the roof of the village hall, causing the walls to collapse. The path of damage extended for about 1·5 kilometres.[16]

Twenty-two tornadoes occurred across southern England on 21 September 1982 in association with a cold front linked to an ex-tropical storm. At Bicester, though the damage was confined to about a 100 metre-wide path, it was considered to be of force 6. Walls were blown over and roofs of warehouses sucked off, and there was also the usual removal of ridge tiles and slates.

The number of tornadoes, reported and verified, for the years 1982–90 is:[17]

Year	1982	1983	1984	1985	1986	1987	1988	1989	1990
Number	76	33	45	20	19	13	14	12	31

(There were nine Severe Tornadoes, TORRO Force 4 or above, in these nine years.)

On 14 December 1989 a man in Long Stratton, Norfolk, looked out of the window and saw his Volvo, without its windows, wheel trim and exhaust, blow past, rolling over and over; then the wind changed direction and it blew back again. Another eye-witness described how he heard a bang in the sky and then the wind vacuumed up all the water in the car park and threw it down again.[18] He saw bricks tumbling and then spotted a metal barrel flying straight at his car. He reclined in his seat just in time as the barrel smashed through the driver's-door window.

David Reynolds, of Tornado Watch[19], forecast the possibility of tornadoes on 11 November 1991 for the next day, and also indicated a likely strength of 4 or 5 on the TORRO Scale. Needless to say, it was not within the bounds of possibility to say exactly where such events would occur. The village of Dullingham, Cambridgeshire, was so afflicted. The Boot public house was seriously damaged and a barn with valuable cars inside was totally destroyed. Other indications of the presence of the tornado were the stripping of ridge tiles and twisted trees. It is probably undesirable for tornado warnings to be given in weather forecasts as the chances of any one place being affected are so small, and panic could occur.

We have all seen a pile of leaves still one moment and in violent rotational agitation the next. Such a miniature whirlwind resembles a tornadic disturbance without the funnel. On a larger scale are the dust devils and water devils of tropical countries. Dust or land devils raise columns of dust measuring from a few metres to a hundred or more high, and traverse the country for a short distance, rotating as they move along, and then ceasing as suddenly as they began. The water devil is their counterpart over the sea or lakes. Much weaker examples are sometimes encountered in Britain (Fig. 53).

On 30 June 1975, a land devil damaged a building at Warmley near Bristol.[20] The day was cloudless and the temperature 22°C (72°F). 'The whirlwind appeared as a dark column of spinning air as it travelled. It lifted a shed off its supports and landed it on a car in the car park about sixty yards away.' Another report, from 1933, comes from a cricketer who played in a match in the Happy Valley, a sheltered downland vale on the east side of Brighton. The weather was hot, still and cloudless. The grass beyond the boundary had been cut for hay and was drying in the sunshine. Suddenly a number of swaths of partly dried grass were lifted high into the air and

Fig. 53. **Water devil on Rydal Water, Lake District.** It moved along at about 16 kilometres (10 miles) an hour, whirring and hissing until it was all of 21 metres (69 feet) high.

deposited some distance away. This time there was no dark spinning column to be seen.[21] In 1976, dust devils were observed in many parts of the country. Several were seen to follow each other across the fields near Cirencester.[22] At Bleasby, Nottinghamshire, 'there was a succession of several during the afternoon. The corn had just been cut and the straw was lifted high into the air, although the day was remarkable for an absence of cloud and wind.'[23]

On 3 July 1977, the weather was cloudless, the maximum temperature was 26°C (79°F), and the wind was a light southerly at Devizes in Wiltshire. In the late afternoon there was 'a shower of hay'.[24] 'A great cloud of it drifted high over the town at teatime. Then, as if dropped on target from a plane, the whole lot fell out of the sky onto the central area of the town. The moonrakers' town suddenly became the haymakers' during the heatwave. The hay came down in the size of handfuls on the light breeze, and blew about the Market Place . . . and adjoining side roads. It was caught in gutters and against buildings and made a big job for the morning street cleaners.' Since the first edition of this book came out in 1982, there have been many other sightings of land devils; but the above examples illustrate their habits.

Over lakes, water devils can raise so much spray or cause such a stir in the water (Fig. 53) that it is easy to understand why some people have assumed them to be animate, particularly as they make a roaring or rushing sound.[25] In the days before there were machines which made such a roaring sound it was naturally assumed that only animate things could do this. Perhaps the occurrence of water devils on Loch Ness could help to explain the mystery of the Monster. Even in ordinary weather the surface of this loch has numerous strange marks on the water caused by the wind. It is, perhaps, sad to think that 'Nessie' may be a water devil.

A party of anglers were fishing on Loch Dionard in Sutherland on 12 February 1978. 'After half an hour or so, a wind blew up and they decided to head for the shore. That is when they heard a whooshing sound behind them. A spout of water, 3 metres high, was bearing down on them. Before they could get out of its path, the boat was lifted out of the water and spun round. It crashed back to the surface facing the other way. The anglers were lucky to get off with nothing worse than a soaking. They watched, terrified but fascinated, as the spout of water swirled across the loch before it fizzled out. When a strong wind blows over the loch and hits the cliffs at a certain angle, it creates a sort of whirlwind that swirls over the surface of the water and sucks up a spout of water. It can prove perilous for the unwary, as the party of anglers found out.'[26] If the event had happened on Loch Ness, it is likely that 'Nessie' would have been held responsible.

NOTES TO CHAPTER 9

1. TORRO, Oxford, and G.T. Meaden, '20 years of tornado research and storm research by TORRO', J.M. 19, pp. 298–303.
2. G.T. Meaden, of TORRO.
3. E.T. Stringer, 'A future problem in urban climatology', University of Birmingham. Paper read at Conference of Geographical Association, 1975.
4. M.W. Rowe, 'Whirlwinds and synoptic patterns', J.M. 2, pp. 197–9.
5. 'The South Wales Tornado of 27 October 1913', Geophysical Mem. No. 11, 1914, Bracknell.
6. A.B. Tinn, 'Tornadoes', This Weather of Ours, Allen and Unwin, 1949, p. 41.
7. L.C.W. Bonacina, 'Tornadoes across England, the Buckinghamshire Tornado', W. 5 pp. 254–5. J. Simmonds, ibid., pp. 255–7. 'Tornadoes in England, 21 May 1950', M.O., Geophysical Mem. No.12, 1957. H. H. Lamb 'Tornadoes in England', M.M. 79, pp. 245–6.
8. F.A. Barnes and C.A.M. King, 'A tornado at Tibshelf, Derbyshire' W. 7, pp. 214–20.
9. S.G. Abbott, 'Thunderstorms and tornadoes of 8 December 1954', W. 10, pp. 142–6. R.E. Lacy, 'Tornadoes in Britain during 1963–66' W. 23, p. 116–24.
10. E.G. Gilbert and J.M. Walker, 'Tornado at the Royal Horticultural Society's garden, Wisley', W. 21, pp. 211–14. J.A. Radford, 'Formation of the Wisley tornado', W. 21, pp. 214–15.
11. W.T. Roach, 'The Barnacle Tornado, 21 April 1968', W. 23, pp. 418–25.
12. P.B. Wright, 'A tornado in South Yorkshire and other tornadoes in Britain', W. 28, pp. 416–28. J.R. Tagg, 'Tornado at Cranfield', W. 29, pp. 60–3. K.W. Whyte, 'Tornado of 26 June 1973', M.M. 103, pp. 160–71.
13. C.E. Biscoe and M. Hunt, 'Severe outbreak of nocturnal tornadoes in East Anglia', J.M. 2, pp. 69–73.
14. R. McLuckie and M.M. Taylor 'The Grantham-Newark tornado of 8 October 1977', J.M. 3, pp. 35–9. 'Eye-witness accounts of the Grantham-Newark tornado', J.M. 3, pp. 40–2. P.S.J. Buller, 'Damage caused by the Grantham-Newark tornado of 3 January 1978', J.M. 3, pp 229–31.
15. G.T. Meaden, 'Tornadoes in Eastern England on 3 January 1978', J.M. 3, pp. 225–9.
16. P.S.J. Buller, 'Damage caused by the Llandissilio tornado, 12 December 1978' J.M. 4, pp. 206–9.
17. See also 'TORRO, British tornadoes 1974–77', J.M. 3, p. 39. 'British tornadoes, annual total 1958–79', J.M. 5, p. 220.
18. 'TORRO Tornado Division Report: November and December 1989', J.M. 15, p. 295.
19. W.S. Pike, 'The two East Anglian tornadoes of 12 November 1991 and their relationship to a fast-moving triple point of a frontal system', J.M.17, pp. 37–49.
20. G.T. Meaden, 'Strength of an English devil', J.M. 1, p. 32.
21. Worman, 'Happy Valley dust devils', John Lyon School, Harrow, 1933.
22. Dr Gould, Faringdon, Berkshire, letter to the author.
23. P. Ripley, Bleasby, Nottinghamshire, letter to the author.
24. Bath and West Evening Chronicle, 'Shower of hay at Devizes', J.M. 2, p. 270.
25. G.T. Meaden, 'A meteorological explanation for some of the mysterious sightings on Loch Ness and other lakes and rivers', J.M. 1, pp. 118–23.
26. 'Loch waterspout upsets anglers', J.M. 3, p.117. J.M. calls it a waterspout rather than a water devil.

10 | Drought

Droughts in the United Kingdom do not pose the very real threat to lives and livelihoods that accompanies periods of severe rainfall deficiency in many parts of the world. Nonetheless, a number of protracted episodes of low rainfall and high temperatures in the 1990s have served to underline our continuing vulnerability to drought conditions. As a consequence, and partly as a result of speculation regarding the impact of climate change, scientific and public interest in drought and its effects is at an unprecedented level. But what is a drought?

Whilst in broad terms the concept of a drought is readily recognised by the public at large, translating this intuitive understanding into an objective procedure for assessing drought severity is far from straightforward. In part, this reflects the difficulties involved in quantifying a phenomenon which varies in its extent, duration and intensity both regionally and locally. For many years the Meteorological Office used to talk of an official drought after no measurable rainfall had fallen for fifteen days. However, 'official' droughts have different consequences in January and July and the impact can depend greatly on the amount of antecedent rainfall and the state of the ground. In recognition of their limited value, the designation of 'official' droughts ended in 1961. The problem of quantifying severity reflects the fact that droughts are multifaceted both in their character and range of impacts. Each drought leaves a unique climatic fingerprint which needs to be effectively documented if meaningful comparisons between drought episodes are to be made (see Fig. 54). The 1975–76 drought, for example, achieved a remarkable intensity in central southern England over a sixteen-month timespan. By contrast, the 1984 drought was largely restricted to the spring and summer and was most severe in the normally wetter northern and western parts of Britain.[1] Both these droughts featured a sequence of dry months bracketed by very wet conditions which clearly delineated the drought's duration. The 1988–92 drought was a more complex event separated into a number of more, or less, severe phases. Though generally less intense than the 1976 drought, the 1988–92 one was remarkably protracted in the eastern lowlands where substantial rainfall deficiencies extended well beyond four years.

Perhaps of equal importance in attempting to assess drought severity are the differing impacts of meteorological droughts, defined essentially on the basis of rainfall deficiency; hydrological droughts, where accumulated shortfalls in runoff and recharge are of primary importance; and agricultural droughts, where the availability of soil water through the growing season is the critical factor. The impact on the community during most periods of large rainfall deficiency is likely to be very uneven and dependent on a number of features of the drought. Hot weather and dry soils may generate heavy water demand in the spring, for irrigation and garden watering in particular. This can overstretch the water-distribution systems and trigger hosepipe bans at a time when overall water resources may be relatively healthy. Conversely a wet summer, as in 1992, may suppress demand and greatly moderate restrictions on water use when resources are at historically very depressed levels.

An additional factor is the public perception of drought severity which may vary considerably from individual to individual. A very hot, dry summer, for instance, is likely to be viewed in a more relaxed manner by the holiday-maker than the farmer or industrialist reliant on river abstractions, especially if reservoir or groundwater stocks are sufficient to provide continuity of domestic supplies. In such circumstances, the environmental stress resulting from drought conditions may be of considerably greater importance than the impact on public water supplies.

So no single methodology for assessing drought severity is likely to be able to accommodate all the relevant variables and to reflect regional and temporal differences in public attitudes both to the impact of drought on the landscape and the aquatic environment and to the inconvenience associated with restrictions on water use introduced to counteract its effects.

Most of this chapter will examine the occurrence and distribution of meteorological droughts, generally limiting the description to droughts of a calendar month or longer, particular attention being directed to the notable drought episodes of the 1990s. Most recent droughts have been protracted affairs but media attention commonly focuses only on the most arid interludes. This can create a misleading impression; for instance, the number of successive dry days is often a poor index of drought severity. Nonetheless, lengthy sequences of rainless days rightly attract interest in a country whose weather conditions are notably capricious.

What is the record for successive days without rain in Britain? In the famous spring drought of 1893, some places in south-east England had no rain for fifty consecutive days, or more. Hailsham, near Eastbourne, had no rain for sixty-one days from 17 March to 16 May; but the longest drought of all was in London, in Mile End, and lasted seventy-three days, from 4 March to 15 May. At Twickenham, the drought was broken a day earlier.[2]

In 1938, a remarkable spring drought of thirty-eight days was recorded at Port William, south-west Scotland, from 3 April to 10 May. Central Ireland also had a prolonged drought. Hull and Scarborough had no rain for forty-one days in February and March 1943. No measurable precipitation fell at Glenquoich in February 1947, though its monthly average is 252 mm (9.9 in). Many parts of west Scotland were so dry that peat fires were observed burning,[3] in sharp contrast to the snow-covered landscapes of the rest of Scotland, and virtually all England and much of Ireland.

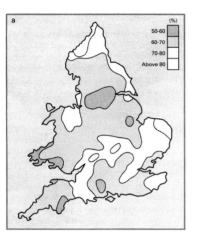

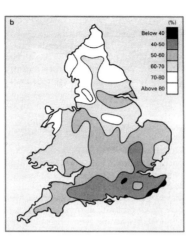

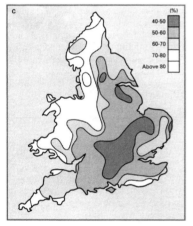

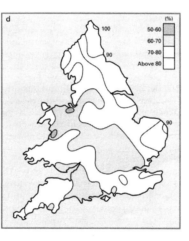

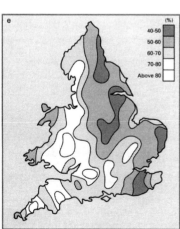

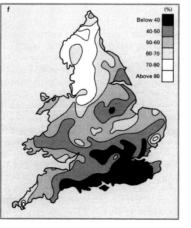

Fig. 54. **Spatial variation in rainfall in England and Wales (% of long-term average) for six drought episodes:**
(a) February-October 1887;
(b) February-November 1921;
(c) February-September 1929;
(d) April-September 1934;
(e) February-September 1959;
(f) January-August 1976.[7]
(Reproduced by kind permission of the Institute of Hydrology)

Cold Arctic or Polar Continental air on the southern side of high pressure over Iceland or Scandinavia covered the greater part of Britain. Because of its origins and low temperatures, the airstream could hold very little moisture. What it did hold was precipitated in the form of instability showers, or orographic snow, over the hills and mountains of the east. When a similar pressure pattern prevailed in January and February 1963, Fort William in the Great Glen recorded only a trace of precipitation on two days. Again, the greater part of Britain was snow-covered, falls exceeding 30 cm (12 in) being widespread. That is the equivalent of about 30 mm (1·2 in) of rain.

A drought of truly Mediterranean proportions lasted throughout August in 1947. The average maximum temperature exceeded 27°C (81°F) in many parts of the south and sunshine was abundant everywhere. Many places, including the isle of Islay, Aberdeen, wide areas of the Scottish Highlands and Southern Uplands, and parts of Cambridgeshire, Kent and Sussex had no measurable rain. High pressure was dominant with light winds from an easterly point.

In 1953, Gorleston, Brigg (Humberside), Tynemouth, Berwick-on-Tweed and several places near Oswestry (Shropshire) recorded droughts of thirty-five or thirty-six days between 20 February and 26 March.

The summer of 1955 gave a thirty-three-day drought at Camborne, Cornwall, from 1 July to 2 August, and a combined July and August total of 9 mm (0·35 in) at a number of places in Devon. Torquay's August rainfall was 2 mm (0·08 in).

In 1957, the Wye valley had forty-one consecutive rainless days ending on 6 May. April 1957 was only the second completely dry calendar month at Ross-on-Wye since records began in 1857 (the other was June 1925). February 1959 was rainless in East Sussex and around Herne Bay in north Kent. At Oxford only 0·2 mm (0·01 in) fell. Kew's total was 2 mm. May 1959 was one of the driest since 1770. In 1959 quite a number of places in eastern England had no rain from 14 August to 10 October. Among them were Finningley, near Doncaster, with a drought of fifty-nine days, and Lowestoft with fifty-seven. March 1961 was rainless in parts of Somerset and Dorset, while rainfall was below average from Shropshire to Kent, though westerly winds blowing round the north side of the anticyclone brought more than average rainfall to the Hebrides and much of northern Scotland, rather like the classic examples of March 1938 and June 1976. Bideford and Cullompton in Devon were without rain in February 1965. Rainfall was under 12 mm (0·5 in) from South Wales to London and in Northern Ireland, but Gorleston and the Yorkshire Moors above Whitby were wetter than average, largely from shower activity on the east wind. The Scottish Highlands had over 300 mm (12 in) of rain in September 1969. By contrast, no rain fell at all in parts of East Anglia. October of the same year was also rainless near Bedford, with totals of 5 mm or less from Bournemouth across to Norfolk.

After a dry March, April 1974 was very dry as well and many new records were broken. The peat bogs of Kinder Scout, widely accepted as the most arduous and watery section of the Pennine Way, were dusty and crumbly. Impressive Kinder Downfall, where the headstreams of the Mersey begin, was contributing no water at all.[4] For most of the month, pressure was high to the north and north-east so that easterly winds prevailed (Fig. 55). The Yorkshire Moors, Cheviots, Lammermuirs and the eastern Highlands (high ground facing the wind) were the only areas to have appreciable rain. The lowlands of Lancashire and Cheshire had no rain at all, along with the west coast of Scotland, a sharp reversal of normal.

Completely rainless calendar months are not uncommon. During the thirty-five years 1940–74, there were eighteen months that were rainless in some part of Britain. February heads the list with four dry months. March comes next with three. April, August and October were rainless twice; and all the other months reported one instance, barring December and January which had none. Many will have heard of the saying 'February fill dike', which suggests that February ought to be a wet month; but it is more likely that the idea originates from the fact that evaporation is low and puddles remain on the ground, particularly if the previous months have been wet.

If droughts of thirty days or more occurring in part of two calendar months are taken into account, droughts are even commoner than the statistics show. However, even if we examine dryness on a calendar month basis, we can still find some startling facts. If we adopt a standard of 5 mm (0·2 in) of rain or less (as does at least one weather station in the Monthly Weather Report)[5] as a criterion for dryness, we find that there were sixty-two very dry months somewhere in Britain in the years 1940–74. November and December had one occasion each. January had two, July three, April, May, August and October six, March , June and September seven, and February ten.

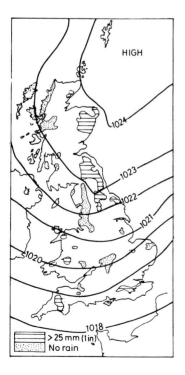

Since 1974, there have been a large number of dry months, too. The three years 1975–77 had ten months when less than 5 mm of rain fell in some part of Britain. They comprised two Julys and two Augusts, and one March, April, June, September, October and December. June 1975 was totally rainless in parts of Wiltshire and at Poole, Dorset, and Ivybridge, Devon. No rain fell at Callington, Cornwall, in April 1976. Dawlish, Devon, and Biddenden, Kent, had none in June 1976. Parts of the Lleyn peninsula were rainless in August 1976. No rain fell at Bardsey Island or Pwllheli. June 1977, the year after the great drought, actually had less rain

Fig. 55. **Rainfall in April 1974**. No rain fell at all in the normally very wet north-west of Scotland. Much of Lancashire and parts of Cumbria were also rainless. In this month, those areas were in the rain shadow of the prevalent wind, which blew from the east. The Cheviot Hills, the Pentland Hills, the eastern slopes of the Pennines and the Yorkshire Moors were the only parts to have significant rain. The isobars show the average pressure for the month.

than June 1976 in several places, but because the month was relatively cool evaporation was low and the effects were much less serious.

The drought of 1975–76 will be remembered for a long time to come. The monthly rainfall totals for Kew for 1975, 1976 and 1995 are given in Table 16 together with the 1941–70 averages, for comparison. Since 1978 there have been fifteen dry months with 5 mm or less of rain. February, May and August had three; June two and March, April, June and October one each.

Many times during the summer of 1976 the whole nation heard at length about the drought in South Wales, southern England and East Anglia. Undoubtedly the tinder-dry state of the ground, the numerous heath and forest fires, the almost total failure of some farm crops, including grass, so that cattle had to be fed on winter fodder in August, and the apparent death of trees such as beech and birch, and of some shrubs, notably rhododendron, indicated the seriousness of the situation. Ponds, wells and springs which had never dried up in living memory did so in 1976. A contributory factor, especially in parts of the English lowlands, was the lowering of water-tables caused by heavy groundwater abstraction—often over many years. The increasingly pervasive effects of land and water management on river flows and groundwater levels has meant that isolating the impact of a meteorological drought can represent a considerable scientific challenge.

Table 16 Rainfall at Kew in mm

	1975	1976	Average (1941–70)	1995
January	84	15	49	121
February	26	21	38	69
March	72	9	38	52
April	37	9	40	15
May	67	22	48	40
June	17	8	47	21
July	26	25	58	29
August	13	13	63	<1
September	120	107	52	84
October	21	95	52	14
November	64	82	62	26
December	28	67	52	93
Year	575	473	599	564

The eleven months October 1975 to August 1976 had 43% of average, 235 mm (9·3 in) instead of 547 mm (21·5 in).

All statistics have to be treated with care and those relating to the drought of the summer of 1976 are no exception. If the drought is measured by calendar year rainfall totals alone, 1976 was not as dry as 1921 (Fig. 56). But the January to December period is an arbitrary timescale within which to examine drought severity. Nationally, the drought began in May 1975 and ended in August 1976; this sixteen-month sequence is the driest in the England and Wales rainfall series—which begins in 1767—by a significant margin. Thus the 1975–76 drought can be truly described as memorable, particularly when other elements are included: persistently high temperatures, drying winds and abundant sunshine, amounting to a daily average of over ten hours a day for two months on end over East Anglia. Much of the southern half of the country looked like the Mediterranean lands and, if all aspects of the weather are taken into account, the drought was the worst on record in modern times and possibly comparable with historical droughts in 1252–53 and 1714–16.

All the same, less rain fell in 1921. At Margate the total for the whole of 1921 was 236 mm (9·3 in), the only occasion when less than 254 mm (10 in) has been recorded in a year. If we take any sequence of twelve months in 1975–76 from anywhere in the country, the lowest rainfall totals were: 318 mm (12·5 in) at Newark to 31 August 1976, and 315 mm (12·4 in) to the end of June 1976 at St. Catherine's Point, the southern tip of the Isle of Wight. Another very dry year was 1933 (Fig. 57). This year, too, beats the driest twelve months in 1975–76, with totals of 282 mm (11·1 in) at Wickham Bishops, Essex, 294 mm (11·6 in) at Maldon, and 306 mm (12 in) at Southminster.

Rainfall deficiencies over the full compass of the 1975–76 drought are unparalleled but the sequences of dry days, though very notable, were not outstanding. The longest period at the time without rain appears to have been forty-five days at Teignmouth, Devon. Many places in the south of England and South Wales had no rain between 20 July and 28 August. At Thorney Island, near Portsmouth, there were forty-four consecutive rainless days. At Plymouth the total was forty-three; Guernsey, Exeter and Oxford forty. At Kew and Guildford the meteorological drought lasted for thirty-eight days; but for only twenty-seven at Liverpool. Further north still the dry weather was not so significant. During the cool summer of 1978, there was a long dry spell in the south. Newport, Gwent, had no rain for thirty-three days, and Brighton for thirty-two up to the last week of July.

Burnley, Lancashire, had no measurable rain for forty-eight successive days during April and May 1980. From 14 April to 20 May there was no rain at Port Erin, Isle of Man, and the combined April and May total was only 5·2 mm (0·2 in). Other low totals in April 1980 included 2 mm (0·08 in) at St. Helier, Jersey; 1·5 mm (0·06 in) at Burnley; 1 mm (0·04 in) at Hawarden, Clwyd, and Accrington, Lancashire. Even the Lake District was dry with only 6 mm (0·24 in) at Coniston. In May 1980 the dry weather continued and Lowestoft recorded 5 mm (0·2 in) and Fair Isle, between Orkney and Shetland, 3·1 mm (0·12 in). For many parts of the south-east the only appreciable rain during these two months fell on 1 April and 31 May. There were fears of a repeat of the 1976-style drought. Heath and forest fires were numerous throughout Britain and some forested areas were

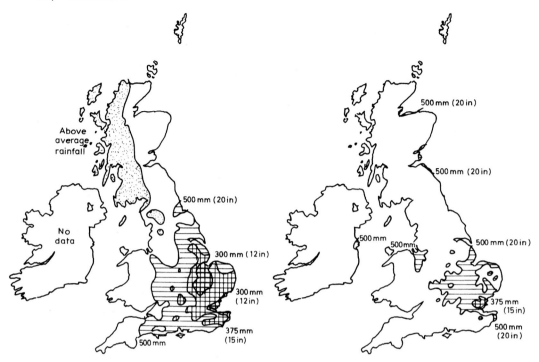

Fig. 56. (*left*) **Rainfall in 1921**. During the whole of 1921, only 236 mm (9·29 in) of rain was recorded at Margate. This is the lowest authentic annual total known in the British Isles. Over western Scotland rainfall was above average for the year.

Fig. 57. (*right*) **Rainfall in 1933**. Virtually all the British Isles had below average rainfall.

closed to the general public to minimise the fire risk. But a very wet spell followed.

We have seen that there are different methods for showing the raininess of a place. So there are for recording dryness.[6] If meteorological drought occurs in the winter, vegetation will not suffer because evaporation will be low. Little moisture is needed by dormant plants. In mid-summer, drought can affect crops even if some rain occurs daily—agricultural drought. In fact, agricultural drought occurs for spells of at least fourteen days in three-quarters of our summers over much of eastern and southern England. This is the reason why fruit and vegetable growers have gone to the expense of installing irrigation equipment. At Kew the worst agricultural drought since the start of records in 1871 occurred in 1959. Next was 1921, the driest year meteorologically, followed by 1893 and 1972 and the three hot summers of 1975, and 1949 and 1911.

For the summer months June, July and August 1976, rainfall was just over a third of normal over England and Wales as a whole compared to 52% in 1949 and 57% in 1921.

Dry winters tend to produce hydrological, or water supply, drought. The effective rainfall for replenishing the reservoirs, filling the wells and keeping the springs and streams flowing is the quantity of rain and snow left to percolate into the ground after evaporation has taken place, and any soil moisture-deficit (S.M.D. or soil dryness) has been removed. It is the steady winter rains which do most of the replenishing. The summer downpours only superficially wet the ground and the rainfall is lost, either by absorption in the soil, or by evaporation (both directly and as a result of transpiration from vegetation). The winter of 1975–76 was not among the ten driest from the water supply angle at Kew, though things were different elsewhere. Probably 'the drought of seventy-six' was made worse by wasteful use of water and the enormous demands of industry. At Kew hydrological drought conditions were more severe in the twelve months beginning July in 1890, 1933, 1964 and 1972. In these four seasons the effective rainfall was virtually nil. There was no significant replenishment of groundwater stocks. Taking all the three types of drought together, 1976 was the worst on record.

February 1986 was a very dry month. Most of the precipitation was in the form of snow and wide areas away from the east coast had negligible totals: 1 mm rain equivalent at Rhoose, Cardiff and Swansea, 1·8 mm at Kendal and even less at Preston and Galway. Durham received 25 mm in what is usually the driest part of the country as it faced the prevailing wind for the month, which came relentlessly from the east, and so collected what moisture there was.

Rainfall for Great Britain over the four-year period beginning in the spring of 1988 was very close to the long-term average but for much of that time many rain-bearing weather systems followed a relatively northerly track remote from the English lowlands.[7] As a result the normal north-west/south-east rainfall gradient across Britain was greatly accentuated, north-west Scotland being very wet whilst eastern and southern England was exceptionally dry. For example, in March 1990 many totals from Essex across to Sussex were below 5 mm (0·2 in), while parts of the Highlands had 273% of average rainfall. This unusual and very persistent disturbance to the normal rainfall distribution, together with the abnormally high temperatures which have characterised much of the recent past (and encouraged high rates of evaporative loss), provide the setting for the drought conditions experienced throughout most of the 1989–92 period. In a few places close to the coast the drought remained severe for most of the four years ending in the summer of 1992. Elsewhere, several wet episodes, notably the winter of 1989–90, served to divide the drought into relatively distinct meteorological phases. A significant reduction in long-term rainfall deficiencies occurred in the first half of 1991, and again in the spring of 1992, but whilst in rainfall terms the drought had lessened, its severity with respect to river runoff and groundwater levels in the east was barely noticeable until the autumn of 1992.

Notable drought conditions characterised parts of the north-east in late 1989 but for England and Wales as a whole, the drought was greatest over the period beginning in March 1990 - which was almost rainless in parts of the south-east. Near-record evaporative losses contributed to the drought's intensification through the summer and, following a relatively dry winter, the water resources outlook remained fragile throughout much of 1991. Notwithstanding a relatively wet spring in 1992, the 28-month rainfall total up to and including June 1992 is eclipsed only by the minima established during the prolonged droughts of the mid–1850s and late 1780s. Over the longest timespans the drought was markedly more severe in the eastern lowlands of England. In some parts of East Anglia, and a few localities to the north, both 1990 and 1991 rank amongst the three driest years this century and accumulated deficiencies over the four years from the spring of 1988 were the equivalent of a full year's rainfall.

A notably wet September in 1992 triggered a further transformation in climatic conditions and the following two and a half years were, in most regions, exceptionally wet and remarkably mild. This very unsettled phase culminated in the winter (December to February) of 1994–95—the wettest on record for Britain as a whole, in a series from 1869. However a marked change in synoptic patterns in the early spring of 1995 heralded a further dramatic transformation. The frequency of westerly and south-westerly airstreams declined markedly through the early spring of 1995 as a northward extension of the Azores high-pressure cell deflected most rain-bearing frontal systems to the north, allowing subtropical air-masses to penetrate across much of the British Isles. Rainfall deficiencies built up quickly through the spring and a heatwave during much of July and August produced a marked intensification in drought conditions. Much of the late spring and summer rainfall in 1995 resulted from patchy showers or localised thunderstorms. Some areas, including parts of West and South Yorkshire, failed to benefit from the highly variable rainfall and experienced particularly intense drought conditions (see Table 18 and Fig. 58). Substantially below-average rainfall was recorded for each of the five months to August 1995 in most regions and accumulated rainfall totals were well below half of the 1961–90 average over wide areas.[8] Conditions were especially arid in the late summer: August rainfall totals were less than 15% of average throughout much of England and a few places in the south-east registered zero monthly totals (e.g. in the Brighton and Eastbourne areas). The mean temperature for August established it as the second warmest month, after July 1983, in the 337-year Central England Temperature (CET) series.[9] For England and Wales, the June-August period in 1995 marginally eclipsed 1976 as the driest summer in the 229-year homogenised England and Wales rainfall series.[10] With Scotland registering its second driest summer on record, the June-August rainfall total for Britain also established a new summer minimum in a series from 1869.

Rainfall deficiencies were even more notable between April and August 1995. Precipitation totals over the five months were below half of the average in most regions with the greatest deficiencies in a broad zone across the greater part of northern England and the English lowlands. Pockets of extreme rainfall deficiency—less than 20% of the 1961–90 average—were located in the north Midlands and Derbyshire.

For England and Wales as a whole, the April-August rainfall total is the lowest for *any* five-month sequence in over 200 years; only during the 1921 drought have five-month rainfall totals approaching the 1976 and 1995 minima been registered (Table 17).

Table 17 **Five-Month Minimum Rainfall Totals for England and Wales, 1800–1995**[8]

Rank	Rainfall (mm)	% of 1800–1995 average	End month and year	
1	149	43·1	August	1995
2	155	44·8	August	1976
3	159	50·7	June	1921
4	184	58·7	June	1938
5	185	56·7	July	1826
6	185	59·0	June	1929
7	186	59·3	June	1887
8	187	52·4	April	1854
9	188	57·6	July	1870
10	191	48·8	March	1858
11	191	52·1	September	1959
12	193	59·1	July	1990

Note: only non-overlapping periods are featured

Although rainfall deficiencies in the English lowlands were similar to those in the north, the water resources outlook was of less immediate concern because groundwater levels in the chalk, England's most important aquifer, remained mostly within the normal range, a consequence of the abundant rainfall throughout the winter of 1994–95.

Early September 1995 witnessed a further marked change in weather patterns with a sequence of active frontal systems sweeping across most regions. Several southern areas recorded more rainfall over the first ten days of September than in the preceding ten weeks and localised flooding was widespread. A repetition of the dramatic end to the droughts of 1976 and 1984 seemed possible as the second driest August on record, for the UK as a whole, was followed in parts of southern England by the second wettest September. This encouraging transformation, and the decline in evaporation demands as the growing season came to an end, greatly eased the water supply stress. However, a number of strategically important reservoir systems, including those in the Pennines and the Lake District, failed to benefit from the early autumn rainfall and, with soils dry in most catchments, the seasonal recovery in runoff and recharge rates was weak and patchy.

Throughout most of England and Wales, October 1995 was relatively dry and remarkably mild, concluding the warmest twelve-month sequence in the entire CET series. The synoptic pattern began to change again in November as a persistent anticyclone to the north of the British Isles allowed airflows from the north-easterly quadrant to become dominant, bringing cold and dry conditions which were to continue through much of the 1995–96 winter. The paucity of rain-bearing frontal systems through the late autumn of 1995 produced a re-intensification in the drought. Particularly severe drought conditions again affected the southern Pennines where for some reservoired catchments the accumulated rainfall deficiencies since March—in a time frame critical for water-resource management—were the highest on record. Stocks in a few West Yorkshire reservoirs fell to below 15% of capacity and tankering was required to maintain supplies.

The drought significantly extended its range during 1996 and, despite a relatively wet autumn, rainfall deficiencies were outstanding throughout much of England and Wales by the year-end. A further intensification followed in early 1997 culminating in a particularly arid fifty-day sequence during March and April when precipitation was largely confined to light showers and fog-drip; many catchments in the south-east recorded less than 5 mm (0·2 in) of rain over this period. The two-year period ending in March 1997 was the driest twenty-four-month sequence in the 221-year national rainfall series and, following two dry winters, groundwater levels were again very depressed over wide areas. Many of the exceptionally low late-spring river flows recorded during the 1976 drought were eclipsed, and the failure of springs and winterbournes produced a major contraction in the stream network. The exceptionally dry soil conditions were also creating real difficulties for the farming community. Fortunately reservoir contents, boosted by heavy rainfall in February, remained within the normal range. Nonetheless, the water industry faced another summer during which balancing the needs of abstractors with those of the environment would represent a considerable challenge.

The cluster of notable rainfall deficiencies, together with an accentuation in both regional and seasonal rainfall contrasts and exceptionally high average temperatures over the recent past, has fuelled speculation that climate change is upon us. However, the natural variability of the UK climate is such that any short-term trends need to be treated with caution. Nonetheless, it may be that the climatic conditions experienced over the last decade—though rare in an historical context—may become rather more familiar in a warmer world (see Chapter 24).

Table 18 **Rainfall at Finningley, Yorkshire, 1995** [11]

Month	Rainfall in mm	% of average
January	83	173
February	51	128
March	33	80
April	14	38
May	31	61
June	32	59
July	18	37
August	8	15

Fig. 58. **Leighton Reservoir, Yorkshire, in the summer of 1995**[12]

NOTES TO CHAPTER 10

1. T. J. Marsh and M.L. Lees, 'The 1984 Drought', Hydrological Data UK series, Institute of Hydrology, 1986, 84 pp.
2. B.R. to 1969 for information on drought. M.W.R. from 1970.
3. *Oban and West Highland Times*, February 1947.
4. Author's observations.
5. M.W.R. and C.O.L.
6. R.G. Tabony, 'Drought classification and a study of droughts at Kew', M.M. 100, pp. 1–10.
7. T.J. Marsh, R.A. Monkhouse, N.W. Arnell, M.L. Lees and N.S. Reynard, 'The 1988–92 Drought', Institute of Hydrology, 1994, 80 pp.
8. T.J. Marsh, 'The 1995 drought—a signal of climatic instability?', *Proceedings of the Institute of Civil Engineers*, Wat. Marit.& Energy, 1996, 118, pp.189–195.
9. G. Manley, 'Central England Temperatures: monthly means 1659 to 1973', *Quarterly Journal of the Royal Meteorological Society*, 1974, pp. 100, 389–405 (updated by the Meteorological Office).
10. T.M. Wigley, T.M. Lough and P.D. Jones, 'Spatial patterns of precipitation in England and Wales and a revised, homogeneous England and Wales precipitation series', *Journal of Climatology*, 1984, vol. 4, pp. 1–25.
11. W. Monthly Summary.
12. *Yorkshire Post*, Leeds.

11 | Temperature

Means, monthly means, how shall we tell your meaning?

Sir Napier Shaw, *Drama of Weather*, 1933

How do the sun's rays warm the air? What happens is that the sun's rays warm the ground and the ground warms the air by contact and convection. The earth's surface is the main source of heat for the air; it absorbs 47% of the incoming solar radiation; a further 17% is absorbed by clouds (though recent research suggests that the amount of radiation absorbed by clouds has been underestimated), water-vapour and dust particles in the air, which contribute to some atmospheric warming. The rest (36%) is reflected back to space. So it is not surprising that the higher one goes in the atmosphere the colder it gets, as the distance from the main heat source, the ground or sea, increases.[1]

When a meteorologist talks of temperature he means the shade temperature of the air. If five people were given a thermometer and asked to place it in the shade and read it at an appointed time, five different readings could result. So many things can cause errors, for example reflections from ground and from buildings. To prevent this and to obtain standardised readings, meteorologists mount their thermometers in a beehive-like box or screen with louvred sides (Figs. 59 and 60) which allow the free passage of air but prevent reflected radiation from affecting the instruments inside. The bulbs of the thermometers must also be mounted a standard height above the ground, 1·22 metres or 4 feet, for differences of 4° to 5°C (7° to 9°F) and more can occur in the lowest few feet of the air on a clear night.

A surface which absorbs all the energy falling on it is called a 'black body'. In turn it re-radiates the energy. Most solids and liquids behave like 'black bodies', though gases such as air do not. The surface of the earth behaves like a black body, gaining heat rapidly during the day and losing it at night, especially if the sky is clear. That helps to explain why the temperature of the air near the ground is more extreme than that at a few metres up, which in its turn is more extreme than the free air at a few hundred metres above.

Average temperatures for each month of the year are obtained by keeping a daily record of the screen temperature at any place for thirty years. At the end of each month the totals of the daily maximum are summed, and then divided by the number of days in the month to give the average daily maximum, or mean maximum. After thirty years the 'standard' average can be obtained. The same procedure is adopted

Fig. 59. (*left*) **The weather enclosure in St. James's Park, London.** This is one of Britain's climatological weather stations. Data is sent to the Met. Office at Bracknell once a month. In the foreground is the Stevenson Screen. The larger screen is designed to house a thermograph (Fig. 101). To the right of the screen lies a recording rain gauge.

Fig. 60. (*above right*) **The interior of the large thermometer screen.** The observer is taking the readings. The two thermometers lying almost horizontally record maximum and minimum temperature. A small index is left at the extreme points. The two vertical thermometers form part of the hygrometer for measuring the humidity of the air. On the left is the dry bulb (an ordinary thermometer), and to the right, an exactly similar instrument known as the wet bulb thermometer. Special tables enable the amount of water vapour to be calculated. (Table 4(c), page 40) As the humidity of the air will sometimes determine whether falling snow will lie, or whether lying snow will thaw, further mention will be made of the wet bulb and humidity (page 183). On the left and right, inside the screen, are recording instruments for humidity and temperature.

Fig. 61. (*right*) **Automatic weather stations are increasing in numbers.** They can be placed in isolated spots and save on observers' wages, too. The temperature probes, for example, are more sensitive as the response time is much shorter than with ordinary mercury in glass thermometers. The data can be fed into a computer so instant access is available. There are numerous anemometers in use to measure wind speed, on the Severn Bridge, for example.

This is an example of such a weather station, as produced by ELE International.

for the minimum. If this process is repeated at a sufficiently large number of weather stations, maps like Figs. 62 to 67, based on internationally agreed thirty-year periods, can be compiled.[2] These maps, though, have the average temperatures reduced to mean sea-level by adding a correction of 0·6°C (1·08°F) for 100 metres (328 feet) of altitude, which is considered the normal lapse rate, or fall of temperature with height. They thus show the temperatures over Britain as if the land were entirely flat like the Fens, and that is misleading for the uninitiated although it does make for simplicity. At Braemar the weather station is 339 metres (1112 feet) above sea-level and the mean January maximum is 3·7°C (38·7°F), but with allowance for altitude the computed maximum becomes 5·7°C (42·3°F). In fact, the reduction of temperature averages to sea-level is reasonable because most people live not very far above sea-level.

Fig. 62 shows that the average daily maximum for January is appreciably higher in the south-west and west than in the east. Night minima (Fig. 63) are also highest in the peninsulas and islands of the west surrounded by warm sea and exposed to the prevailing wind. Even on this scale the warmth of London is noticeable.

In July (Fig. 65) the daily maximum is highest in a broad area from the Vale of Evesham to the Thames Estuary and East Anglia, with the highest values occurring in London and the Home Counties. A notable feature is the sharp temperature gradient along parts of the east and south coasts, the effect of cooling sea breezes, which bring it about that Glasgow has warmer afternoons than Portland Bill in Dorset. At night southern coasts are warmest and inland areas relatively cool (Fig. 66).

Climatological maps in school atlases and textbooks invariably use mean temperature. Holiday brochures sometimes give afternoon maxima. Mean values are obtained by adding the average maximum and average minimum together and dividing the result by two (Figs. 64 and 67). Once again, the advantage of this method is its simplicity as only one set of values needs be shown on a map; but two places with the same mean temperature will not necessarily be considered as warm or as cold as each other, nor even have a similar climate, hence the comment at the head of this chapter.

The January means for Scilly and for Nice in the south of France are almost the same; yet most people would expect Nice to be warmer. Indeed, many take winter holidays there because of the favourable climate. In reality Nice is warmer if we take afternoon temperatures, when people are out and about, and when winds are gentle and sunshine strong. On the other hand frost and snow are more likely at Nice than Scilly.

Fig. 64 shows that the Outer Hebrides have the same mean temperature as the south coast of England during January. If, however, a resident of Worthing moved to the Hebrides, he would in all probability feel much colder in those northern isles because the winds are stronger there and calm sunny days are less common. In summer (Fig. 67) the conditions more closely resemble the popular idea of temperature distribution, although the south coast is not the warmest part, London and the Thames valley are! Cornwall is similar to Yorkshire and the city of York has the same mean temperature as Scilly, but York has warmer afternoons and cooler nights.

Average temperature is not a fixed value and averages are revised every thirty

Fig. 62. **Mean daily maximum for January in °C**

Fig. 63. **Mean daily minimum for January in °C**

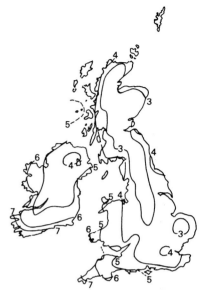

Fig. 64. **Mean daily temperature for January in °C**

Fig. 65. **Mean daily maximum for July in °C**

Fig. 66. **Mean daily minimum for July in °C**

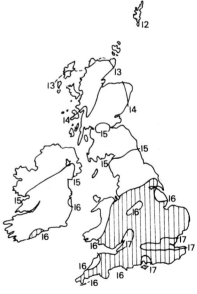

Fig. 67. **Mean daily temperature for July in °C**

years, though variations are usually small between one average period and another, as we shall see in Chapter 24, 'Is Our Climate Changing?' So far we have taken no account of the height of the land. An actual temperature map (for January) would show the necessary allowances for altitude, and looks rather like a relief map. The shapes of the Cotswold Hills and Dartmoor would be shown on such a map by the isotherm for 3°C (37°F). Quite small departures from the average can make an appreciable difference to the winter weather. Over England there have been eight Januarys this century more than 2°C (3·6°F) below average. The effect of this is to increase the likely snow-cover from six to sixteen days.[3] A summer month more than 2°C below average may have less obvious consequences, but it can be serious for the hill farmer as the yields from his land will be lower.

Of statistical interest is the mean annual temperature. While these figures would probably have considerable value for studying long-term climatic change, they tell us little about the year's weather. The figures are obtained by summing the monthly means and dividing by twelve to get the year's average, then adding the yearly totals and dividing by thirty to get the mean for the standard period. Penzance and the Scilly Isles have the highest annual means with 11°C (52°F), and the top of Ben Nevis with just under 0°C the lowest, using data from around the turn of this century. Braemar has the lowest mean, 6·4°C (43·5°F) of any inhabited place where records are kept. Over a number of years the annual means are unlikely to vary by more than 2·5°C (4·5°F) either way. Some of the years with severe winters, such as 1940 and 1947, had good summers to compensate; 1975 had a very mild winter and a hot summer but a notably cool spring. One of the coldest years this century was 1962. The cool summer of 1962 preceded a very severe winter. At Oxford, 1921 and 1949 had means of over 11°C (52°F); 1919 and 1962 were the coldest with means of 8·7 and 8·8°C (47·7 and 47·8°F) respectively. Using various records from the Midlands as far back as 1659,[4] the warmest year was 1949 with a mean of 10·6°C (51·1°F), followed by 1959 with 10·5°C (50·9°F), compared to the average of 9·6°C (49·3°F). One of the few years where the temperatures deviated more than 3·6° to 9°F from average was 1740, with a mean of 6·8°C (44·2°F), similar to Braemar's mean today. The mean was 9·9°C (49·8°F) in 1976, 9·8°C (49·6°F) in 1983, and 10·6°C (51·1°F) in 1990, just fractionally above that in 1949. 1995 may prove to have been the warmest.

So far three basic temperature points have been described: the mean or average maximum, the mean minimum and the mean. There are four other points in the temperature scale: the average warmest day and coldest night of each month and the absolute or extreme maxima and minima. These seven temperature points are set out (Table 19) for six places:[5] Braemar, typical of our coldest inhabited climate; Edinburgh, much influenced by weather from the North Sea; Armagh, typical of inland Ulster; Cambridge, in the more extreme part of lowland England; Kew, typical of a suburban area; and Falmouth in the mild south-west.

It is not unusual to hear people say that a mid-winter afternoon with the thermometer standing at 10°C (50°F) is way above normal. True enough, 10°C is way above average, but such temperatures do occur in the three winter months. Indeed, a

winter month when the temperature does not rise to 10°C is definitely more abnormal. Between 1901 and 1994 the absolute maximum at Nottingham was under 10°C
in only seven Januarys, those of 1940, 1941, 1959, 1963, 1978, 1979 and 1987. In
February, 10°C was not reached nine times, in 1901, 1906, 1930, 1942, 1947, 1963, 1969,
1972 and 1986. The lowest monthly absolute maximum was 5·0°C (41°F) in January
1963 and 4·2°C (39·6°F) in February 1986. At the other extreme 15°C (59°F) occurred
in eight Februarys. Generally, 10°C will be reached at least once at all lowland stations
in an average winter month, and 13°C (55°F) is quite the rule in the lee (east) of high
ground, where a föhn effect operates (see Chapter 20). Not far from the shores of the
Irish Sea, at Bidston Observatory[6] on the Wirral, there have been eleven Januarys
since 1867 when the thermometer has failed to reach 10°C; and eighteen Februarys in
127 years. In February 1986 the month's absolute maximum was 3·9°C (39°F).

At St. Andrews, in eastern Scotland, only seven Januarys out of sixty had absolute
maxima below 10°C. On the other side of Scotland, Tiree had five Januarys when the
thermometer failed to reach 10°C over a record of forty-six years.

The average coldest night of the year map (Fig. 68) is much simplified. The lowest
temperatures usually occur on calm clear nights, when the cold air collects quietly
in the hollows, as a result of both radiative cooling and katabatic (or downslope)
flow of cold air from neighbouring hills. Then marked temperature variations are
widespread. Differences as great as 10°C in 100 metres (5·5°F in 100 feet) of altitude
are common in such conditions in undulating country. All the same, the map shows

Fig. 68. **Average coldest night of the year**

Table 19 Some Temperature Standards (°C)

	Absolute maximum	Average warmest day	Average maximum	Mean	Average minimum	Average coldest night	Absolute minimum
BRAEMAR (339 m)							
January	13·3	8·9	3·5	0·4	-2·7	-13·3	-27·2
February	13·9	8·7	3·8	0·5	-2·8	-13·3	-27·2
March	20·6	13·0	6·4	2·7	-0·9	-9·6	-21·7
April	21·1	16·2	9·4	5·2	1·0	-6·0	-13·3
May	27·2	21·1	12·9	8·2	3·5	-2·9	-8·9
June	27·8	23·7	16·3	11·5	6·7	0·5	-3·3
July	30·0	23·3	17·1	12·7	8·3	2·5	-1·7
August	30·0	22·7	16·6	12·3	8·0	0·9	-2·2
September	23·9	20·1	14·3	10·3	6·4	-0·8	-6·1
October	23·0	16·5	11·6	7·8	4·0	-4·0	-10·6
November	15·6	11·6	6·5	3·4	0·3	-9·0	-23·3
December	12·8	9·9	4·5	1·7	-1·1	-10·8	-21·7
Year	30·0	25·4	10·2	6·4	2·6	-17·1	-27·2
EDINBURGH, ROYAL BOTANIC GARDEN (26 m)							
January	14·4	11·6	5·9	2·9	0·0	-7·1	-15·6
February	15·0	11·5	6·6	3·4	0·2	-6·5	-11·7
March	22·2	14·3	8·7	5·3	1·8	-4·4	-11·7
April	21·7	17·8	11·7	7·7	3·7	-2·4	-6·1
May	24·4	20·7	14·2	10·1	6·0	0·6	-1·7
June	29·4	23·5	17·3	13·2	9·1	4·6	1·5
July	28·3	23·5	18·6	14·7	10·8	6·2	4·4
August	30·0	23·2	18·2	14·4	10·6	5·8	2·2
September	24·4	21·4	16·6	12·8	9·0	2·8	-2·2
October	24·4	18·2	13·3	9·9	6·5	-0·7	-3·7
November	20·6	14·3	9·0	5·8	2·6	-3·8	-8·9
December	16·1	12·3	7·0	4·1	1·1	-5·9	-11·1
Year	30·0	25·3	12·3	8·7	5·1	-8·8	-15·6

Table 19 **Some Temperature Standards** (°C) *continued*

	Absolute maximum	Average warmest day	Average maximum	Mean	Average minimum	Average coldest night	Absolute minimum
ARMAGH (62 m)							
January	13·9	11·6	6·7	3·9	1·2	−5·6	−11·1
February	14·4	11·8	7·4	4·3	1·2	−4·4	−10·6
March	21·7	15·6	10·0	6·3	2·6	−3·5	−12·2
April	21·7	18·0	12·7	8·3	4·0	−1·3	−4·4
May	25·6	21·7	15·5	10·9	6·3	−0·7	−1·7
June	30·0	24·3	18·3	13·7	9·2	4·3	0·6
July	28·9	24·0	19·0	14·9	10·8	6·4	4·4
August	29·0	23·1	18·8	14·7	10·6	5·6	2·2
September	26·7	21·2	16·8	13·0	9·2	3·6	−0·6
October	21·7	18·0	13·6	10·3	6·9	0·7	−2·2
November	16·7	14·0	9·5	6·5	3·5	−2·9	−6·7
December	15·0	12·2	7·6	4·9	2·1	−4·4	−9·5
Year	30·0	25·7	13·0	9·3	5·6	−7·1	−12·2
CAMBRIDGE (12 m)							
January	14·5	12·1	6·0	3·2	0·4	−7·4	−16·1
February	17·8	12·8	6·9	3·7	0·5	−6·2	−17·2
March	23·9	17·3	10·1	5·9	1·6	−4·9	−11·7
April	26·1	20·5	13·7	8·9	4·0	−2·2	−6·1
May	31·1	25·0	17·2	11·9	6·7	0·2	−4·4
June	34·0	27·0	20·4	15·1	9·7	3·9	−0·6
July	34·4	28·4	21·7	16·7	11·7	6·4	2·2
August	35·6	27·8	21·4	16·4	11·4	6·1	3·3
September	33·9	25·0	19·1	14·4	9·7	2·6	−2·2
October	26·7	20·7	15·0	10·9	6·7	−0·4	−6·1
November	21·1	15·2	9·8	6·5	3·3	−3·7	−13·3
December	15·8	12·7	7·1	4·2	1·3	−6·0	−15·6
Year	35·6	30·1	14·0	9·8	5·6	−9·5	−17·2

Table 19 **Some Temperature Standards** (°C) continued

	Absolute maximum	Average warmest day	Average maximum	Mean	Average minimum	Average coldest night	Absolute minimum
KEW OBSERVATORY, RICHMOND (5 m), 1866–1980, AND ROYAL BOTANIC GARDEN, KEW (6 m), 1981–95							
January	14·4	11·8	6·1	4·2	2·3	-4·5	-12·8
February	16·7	12·2	6·8	4·5	2·3	-3·4	-11·7
March	22·0	16·5	9·8	6·6	3·4	-2·2	-8·3
April	26·7	20·1	13·3	9·3	5·3	0·6	-2·1
May	30·6	24·3	16·8	12·6	8·4	3·0	-3·1
June	34·2	26·4	20·2	15·9	11·5	7·0	-0·6
July	33·9	28·4	21·6	17·5	13·4	9·3	3·9
August	36·2	26·9	21·0	17·1	13·1	8·5	2·1
September	33·3	23·7	18·5	14·9	11·4	5·7	1·4
October	27·8	19·9	14·7	11·6	8·5	1·4	-3·9
November	18·9	14·9	9·8	7·5	5·3	-1·6	-7·1
December	15·7	12·3	7·2	5·3	3·4	-3·5	-11·7
Year	36·2	29·7	13·8	10·6	7·4	-5·8	-12·8
FALMOUTH (51 m)							
January	13·9	12·3	8·9	6·5	4·0	-2·2	-8·7
February	14·4	12·4	8·5	6·0	3·5	-2·0	-7·8
March	17·2	14·0	10·2	7·6	5·0	-0·4	-6·7
April	22·8	16·8	12·5	9·4	6·3	1·9	-1·7
May	26·7	19·9	14·9	11·7	8·4	3·7	0·6
June	29·0	22·9	17·8	14·5	11·1	7·1	4·4
July	29·4	23·5	19·2	15·9	12·7	9·2	6·7
August	29·4	23·0	19·2	16·1	12·9	9·5	6·7
September	25·6	21·0	17·5	14·6	11·7	7·1	2·2
October	22·2	18·1	14·9	12·3	9·7	4·4	0·0
November	17·5	12·9	11·7	9·1	6·5	1·0	-5·0
December	15·0	12·8	9·9	7·5	5·0	-0·5	-5·6
Year	29·4	24·6	13·8	10·9	8·1	-4·0	-8·7

The averages refer to the standard period 1941–70. Absolute extremes are correct to August 1995.

the generally expected minima in open country. Anything lower than −4°C (25°F) is uncommon round the coasts, particulary where winds blow off the sea. Central London is also free of anything other than slight frost in most winters, as is much of the south and west coast. Almost one January in six is frost free at Tiree, although the Scillies probably hold the record for freedom from frost (Table 26). Of all the weather stations in the United Kingdom, Scilly has the highest absolute minimum of −7·2°C (19°F). That is not surprising as it is 50 kilometres (30 miles) from Land's End and surrounded by a warm sea, the temperature of which is unlikely to fall much below 10°C even in a severe winter. On the mainland, St. Ann's Head and Dale Fort in south-west Wales probably have the highest minima of −7·2°C (19°F), followed by Falmouth with −7·8°C (18°F). It appears that west Wales is just too far south to receive the worst of the cold from Arctic air and just too far west to experience severe frost in Polar Continental air.

The average warmest day of the year map (Fig. 69) shows that 27°C (81°F) is exceeded over all of England from the Vale of York to the south coast, including the East Anglian and Kent resorts. Most of Scotland exceeds 24°C (75°F), as does most of Wales, Devon and east Cornwall. In the Inner Hebrides, Skye and Rum, the warmest day will be about 21°C (70°F) and only a degree or two higher at Scilly. By the standards of the south of England, there will be no warm days in an average summer in the Orkneys and Shetlands. The absence of very hot days at Scilly evokes the question: 'But what about the sub-tropical gardens at Tresco? Surely these exotic palms need great heat?' One might also ask: 'What about the palms and fuchsias at Inverary Castle, in west Scotland?' There the summers are as cool as anywhere else on the mainland, except for the mountain tops. The paradox is simply explained; the tropical plants flourish in these latitudes because there is usually so little frost in winter. No amount of summer warmth can compensate for one killing frost.

Maximum temperatures likely once in fifty years are shown in Fig. 70. Almost all of England and Wales and the greater part of Scotland below 200 metres (656 feet) have reached 30°C (86°F). Only the Scilly Isles and St. David's Head have not experienced such great heat. In the Outer Hebrides the record is 25°C (77°F), a value also reached in a sheltered part of the Shetland Isles. From the south coast to Yorkshire many places have recorded 34° or 35°C (93° or 95°F), while 32·2°C (90°F) or more has been recorded at the coastal resorts from Bournemouth round to Scarborough and at Rhyl, Aberystwyth, Southport and Blackpool. In Scotland, Perth and Kilmarnock have reached 32°C as well. In Ireland, the record stands at 33·4°C (92·1°F) in Dublin. Kilkenny and Limerick have also reached 32·2°C (90°F). In fact, a reading of 32°C is not so uncommon in the British Isles, occurring somewhere one year in two.

Fig. 71 shows the lowest temperatures likely once in fifty years, in fairly level open country. Some extremes for recent years are given in Table 20.[7] Details of the highest and lowest temperatures are shown in Table 21.[8]

We have seen that the thermometer will exceed 27°C (81°F) over much of the mainland in an average year. But how often is there a hot day? Taking 25°C (77°F) as the

Table 20 Annual Extremes of Temperature for England (°C)

		Highest		*Lowest*
1939	32·2	London	−15·0	Newport, Shropshire
1940	32·7	Cranwell, Lincs	−21·1	Bodiam, East Sussex; Ambleside
1941	34·4	London	−21·1	Houghall, Durham
1942	33·8	Sprowston, Norfolk	−15·5	Woburn, Bedfordshire
1943	33·8	Worcester; Croydon	−10·5	Buxton, Derbyshire
1944	32·7	London; Tunbridge Wells; Horsham	−12·7	Belper; Walsall; Newport, Shropshire
1945	32·2	Norwich; Whitstable	−18·3	Shobdon, Hereford
1946	30·5	London; Finningley (Doncaster)	−13·3	South Farnborough; Woburn
1947	34·4	London; Waddington	−21·1	Houghall
1948	34·4	London; Mildenhall	−15·0	East Malling
1949	33·3	Worcester	−9·4	Farnham; Goudhurst; Houghall
1950	33·3	London	−14·4	Droitwich
1951	30·0	Southend	−13·3	Houghall
1952	33·3	London	−14·4	Shawbury
1953	33·8	London	−11·6	Houghall
1954	30·5	London	−17·2	Haydon Bridge, Northumberland
1955	32·2	London; Chivenor, Devon	−18·3	Houghall
1956	30·3	London	−17·2	Haydon Bridge
1957	35·5	London	−16·1	Sutton Bonnington, Nottinghamshire
1958	29·4	Leicester	−19·4	Shawbury
1959	35·5	Gunby, Lincolnshire	−13·8	Moor House (Pennines)
1960	30·5	Wyton, Cambridgeshire	−13·8	Kielder Castle
1961	33·9	London; Dartford, Kent	−17·2	Kielder Castle
1962	27·8	Writtle, Essex	−16·0	Elmdon
1963	28·9	Wisley; Littlehampton	−20·6	Hereford; Stanstead Abbots, Herts
1964	32·8	Cromer	−15·5	Caldecott, Leicestershire
1965	28·9	London	−20·0	Houghall
1966	28·9	London; Southampton	−18·9	Elmstone, Kent
1967	30·2	Watnall, Nottinghamshire	−13·2	Wallingford, Oxfordshire
1968	33·3	London	−14·4	Cromer
1969	32·8	Letchworth, Hertfordshire	−20·0	Newton Rigg, Cumbria
1970	32·2	Aldenham; Stratford-on-Avon	−16·7	Newton Rigg
1971	30·3	Rugby	−12·4	Moor House
1972	28·5	Southampton	−18·5	Moor House
1973	32·1	Southampton	−15·8	Moor House
1974	27·0	Stratford-on-Avon	−8·4	Santon Downham, Norfolk

Table 20 **Annual Extremes of Temperature for England (°C)** *continued*

	Highest		*Lowest*	
1975	34·2	London (Heathrow)	−10·1	Easthampstead, Berkshire
1976	35·6	Cheltenham; Plumpton, East Sussex	−16·0	Grendon Underwood, Oxfordshire
1977	29·1	Carlisle	−16·4	Moor House
1978	28·4	March, Cambridgeshire	−15·1	Moor House
1979	30·6	London Weather Centre	−16·8	Woburn
1980	29·2	Aldenham, Hertfordshire	−10·9	Moor House; Newton Rigg
1981	29·7	London Weather Centre	−25·2	Shawbury
1982	30·5	East Bergholt, Suffolk	−26·1	Newport, Shropshire
1983	31·7	Malvern; Heathrow	−10·2	Kielder Castle
1984	31·7	Heathrow	−11·1	Widdybank Fell, Pennines
1985	30·5	Mepal, Cambridgeshire	−19·5	Caldecott, Oxfordshire
1986	32·0	Rustington, West Sussex	−18·9	Elmstone, Kent
1987	30·1	Northolt	−23·3	Caldecott, Oxfordshire
1988	30·2	Cheltenham	−10·4	Hartburn Grange, Cleveland
1989	34·0	Mepal	−10·3	Grendon Underwood
1990	37·1	Cheltenham	−9·0	Grendon Underwood
1991	32·1	Cromer	−16·0	Cawood, York
1992	30·3	Malvern	−12·3	Barbourne, Hereford & Worcs.
1993	29·7	East Bergholt	−12·6	Haydon Bridge, Northumberland

Table 21 **Some Temperature Extremes**

Start (and end) of record		Maximum (°C)	Minimum (°C)
	LONDON AREA		
1910	Hampstead	34·6	−12·8
1903	St. James's Park	35·6	−10·0
1948	London Airport (Heathrow)	36·5	−13·2
1871–1980	Kew Observatory	34·4	−12·8
1981	Kew Gardens	36·2	−10·9
	SURREY		
1904	Wisley (R.H.S.)	35·6	−15·1
	KENT		
1934	Goudhurst	34·7	−19·4
1901	Dungeness	30·6	−13·3
1907	Dover	34·2	−15·4
1901	Margate (except 1940–43)	34·4	−8·3
	SUSSEX		
1901	Eastbourne	32·6	−11·1
1901–80	Brighton	33·3	−10·0
1901	Worthing	33·1	−10·6
	HAMPSHIRE AND THE ISLE OF WIGHT		
1901	Southampton	35·6	−11·7
1901	Portsmouth	32·8	−11·1
1925	St. Catherine's Point	30·0	−9·5
1907	Sandown	32·0	−10·6
	WILTSHIRE		
1901	Marlborough	35·6	−19·0
	SOMERSET		
1928	Cannington (near Bridgwater)	34·0	−15·6
	DORSET		
1901	Bournemouth	34·1	−11·1
1901	Weymouth	32·3	−9·5
1901	Portland Bill	30·8	−10·4

Table 21 **Some Temperature Extremes** *continued*

Start (and end) of record		*Maximum (°C)*	*Minimum (°C)*
	DEVON		
1901	Plymouth Hoe	30·6	−8·9
1911	Tavistock	31·7	−16·6
1901	Princetown (except 1965–73)	31·4	−13·3
1901	Torbay	31·0	−8·9
	CORNWALL		
1913	Bude	32·2	−11·1
1901	Penzance	29·9	−9·9
1901	St. Mary's (Scilly Isles)	27·8	−7·2
	CHANNEL ISLES		
1949	Guernsey	32·8	−7·8
1949	Alderney	29·3	−8·9
1901–82	Jersey (except 1941–45)	35·6	−8·3
1951	Jersey Airport	34·5	−10·9
	AVON		
1920	Long Ashton	33·9	−14·4
	GLOUCESTERSHIRE		
1901	Cheltenham	37·1	−20·1
	HEREFORD AND WORCESTER		
1901–50, 1962–74	Hereford	33·3	−20·0
	OXFORDSHIRE		
1871	Oxford (Radcliffe)	35·1	−16·6
	WARWICKSHIRE		
1933	Stratford-upon-Avon	35·7	−21·0
	WEST MIDLANDS		
1901	Birmingham (Edgbaston)	34·8	−12·9
1949	Elmdon	34·9	−20·8

Table 21 **Some Temperature Extremes** *continued*

Start (and end) of record		Maximum (°C)	Minimum (°C)
	HERTFORDSHIRE		
1901	Rothamsted	33·8	−17·0
	BEDFORDSHIRE		
1901	Woburn	34·4	−20·6
	ESSEX		
1941	Writtle (Chelmsford)	35·2	−20·6
1903	Shoeburyness	33·3	−12·8
	SUFFOLK		
1871	Gorleston	31·7	−12·0
	LINCOLNSHIRE		
1904	Skegness	32·4	−13·3
	NOTTINGHAMSHIRE		
1871	Nottingham	34·4	−18·3
	YORKSHIRE & HUMBERSIDE		
1901	Hull	34·4	−11·1
1880	Spurn Head	30·6	−10·6
1901	York	33·3	−13·9
1872	Scarborough	32·8	−13·3
1901	Harrogate	33·1	−13·3
1950	Malham Tarn	28·2	−11·1
1901	Sheffield	34·3	−11·7
1907–82	Huddersfield (Oakes)	31·7	−11·7
1908	Bradford	32·2	−13·9
	DERBYSHIRE		
1901	Buxton	32·7	−18·3
	CHESHIRE		
1901	Macclesfield	33·1	−13·9
	MERSEYSIDE		
1876	Bidston Observatory	34·5	−10·7
1901	Southport	33·9	−16·1

Table 21 **Some Temperature Extremes** *continued*

Start (and end) of record		Maximum (°C)	Minimum (°C)
	GREATER MANCHESTER		
1908	Bolton	32·1	−13·9
	LANCASHIRE		
1915	Morecambe	32·7	−13·3
	CUMBRIA		
1901	Newton Rigg	33·3	−20·0
	CO. DURHAM		
1901	Durham	32·5	−16·1
	TYNE & WEAR		
1912	Tynemouth	31·9	−11·6
	SCOTLAND		
1939	Lerwick	23·4	−8·9
1974	Fair Isle	20·2	−5·1
1901	Kirkwall	25·6	−10·5
1940	Cape Wrath	27·2	−12·0
1871	Stornoway	25·6	−12·2
1923–81	Achnashellach	31·0	−16·1
1947	Benbecula	27·2	−10·5
1923	Onich	30·0	−14·0
1901	Fort Augustus	30·5	−18·3
1931	Dalwhinnie	30·0	−20·5
1901	Inverness	30·5	−12·8
1963–72	Cairngorm	21·4	−15·6
1901	Nairn	30·5	−16·6
1958	Grantown-on-Spey	30·0	−26·8
1901	Braemar	30·0	−27·2*
1871	Aberdeen (Dyce)	28·3	−19·3
1939	Edinburgh (Royal Botanic Garden)	30·0	−15·5
1913	Glasgow (Springburn) (except 1921–37)	31·1	−15·8
1901	Rothesay	28·9	−11·1
1927	Tiree	26·1	−7·0
1910	Eskdalemuir	29·8	−19·0

Table 21 **Some Temperature Extremes** *continued*

Start (and end) of record		Maximum (°C)	Minimum (°C)
	WALES		
1945	Holyhead	32·8	−12·2
1932	Prestatyn	34·3	−13·5
1958	Corwen	31·5	−22·7
1901	Aberystwyth	32·8	−10·6
1936	Llandrindod Wells (except 1941–46)	30·6	−21·7
1908	Swansea	31·7	−10·0
1904	Cardiff	33·5	−16·7
1924	Usk	33·3	−18·3
	NORTHERN IRELAND		
1926	Aldergrove	29·4	−12·7
1901	Armagh	31·1	−11·1
	REPUBLIC OF IRELAND		
1885	Malin Head, Donegal	28·9	−6·7
1938	Shannon Airport	31·6	−11·9
1881	Roche's Point, Co. Cork	28·9	−6·7
1939	Dublin Airport	27·3	−10·9
1892	Valentia	29·8	−7·3
1872	Birr Castle	31·7	−15·6

* −27·2 in February 1895 and January 1982.

standard for a hot (or summer) day, Table 22 [9] shows that it is only in the east and south, away from the sea, that this value will be reached ten or more times a year. At Gorleston, even the best year for hot days when there were nine, is still not up to the average of twelve for Cambridge, and far below Cambridge's highest total of thirty-five days. Southampton's total of forty-two is noteworthy. It is the equivalent of six weeks of hot weather, totally different from the 'three fine days and a thunderstorm' so often derisively alleged to make up our summers. Many ordinary days also just fail to reach 25°C, while 20°C without much wind and with some sunshine can seem quite warm enough to deserve classification as a summer day.

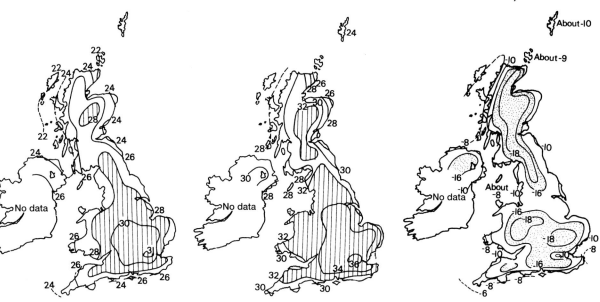

Fig. 69. **Average warmest day of the year (°C)**

Fig. 70. **Maximum temperatures likely once in fifty years (°C)**

Fig. 71. **Minimum temperatures likely once in fifty years (°C)**

Table 22 **Number of Days over 25°C (77°F) During the Periods 1941–70 and 1971–90**

| | 1941–70 | | 1971–90 | |
	Average in a year	*Most in a year*	*Average in a year*	*Most in a year*
Stornoway	0	0	0	0
Aberdeen	0·3	2	0·7	2
Rothesay	1	6	2	9
Renfrew	2·3	13	n/a	n/a
Cockle Park (Northumberland)	0·7	5	1·4	8
Gorleston	1·8	9	n/a	n/a
Cambridge	13·2	37	16·8*	45*
Birmingham (Edgbaston)	8·5	24	11·4	34
Kew	11·4	32	n/a	n/a
Hastings or Lympne	4·3	28	5·2	19
Southampton	11·7	42	14	43
Falmouth	1	7	n/a	n/a
Valley, Anglesey	1·4	7	3·9	19
Bidston	3·7	23	n/a	n/a
Stonyhurst, Lancashire	2·7	13	n/a	n/a
Douglas, Isle of Man	0·3	3	0·8	9
Aldergrove	0·9	6	1·6	11
Armagh	1·7	11	4	18

In 1962 none of the places listed above reached 25°C.

*No data is available for 1984–86.

Table 23 Days Between 20° and 25°C (68° and 77°F) and More Than 25°C in Eight Hot Summers

	1947		1949		1959		1975		1976		1983		1989		1990	
	20–25°C	>25°	20–25°C	>25°	20–25°C	>25°	20–25°C	>25°	20–25°C	>25°	20–25°C	>25°	20–25°C	>25°	20–25°C	>25°
Lerwick	0	0	0	0	0	0	1	0	0	0	0	0	1	0	0	0
Stornoway	15	0	5	0	3	0	6	0	14	0	8	0	4	0	8	0
Craibstone	10	0	21	0	31	2	26	2	34	2	28	2	24	1	17	2
Rothesay	29	2	28	2	30	4	30	4	27	9	30	3	20	4	22	0
Eskdalemuir	38	1	31	0	38	2	31	10	27	11	36	8	19	11	23	3
Cockle Park	52	0	40	5	44	5	30*	7	36	8	15	2			25	3
Gorleston	55	0	48	6	40	4			66	3	35	7	63	10	47	14
Cambridge					84	37	61	29	52	45	43	2	62	41		
Birmingham	53	24	46	24	67	15	49	21	42	34	48	22	60	25	51	26
Kew Observatory	57	32	74	27	80	29	58	27	52	51	46	43	61	48	70	40
Lympne or Hastings	67	28	67	14	80	9	47	11	59	19	53	12	76	8	48	8
Southampton	59	42	85	32	103	33	60	35	53	43	45	37	69	38	65	22
Falmouth	45	5	71	7	49	0	41	2	53	4	40	4	68	5		
Valley, Anglesey	28	0	23	2	45	3	26	4	21	19	35	9	30	8	24	9
Bidston, Merseyside	6	15	46	6	48	4	35	0	40	20	34	6			31	10
Stonyhurst	40	13	45	4	57	5	35	15	34	23						
Douglas, Isle of Man	28	3	24	0	20	0	28	2	19	9	28	2	20	0	17	1
Aldergrove, Belfast	35	5	39	4	44	1	50	3	31	11			32	6	29	2
Armagh	34	9	41	9	58	5	52	9	30	14	47	4	33	18	32	4

* Acklington

Over ten recent years, 20°C was reached or exceeded on an average of seventy days at Kew and twenty at Bidston, Merseyside. Even in the cold summer of 1962, there were forty-eight days when the thermometer reached 20°C, or seven weeks of warm days, though they were not all consecutive, of course. Data for eight recent hot summers are given in Table 23.[10]

NOTES TO CHAPTER 11

1. R.G. Barry and R.J. Chorley, *Atmosphere, Weather and Climate*, Methuen, 1968, p. 33.
2. M.O., Climatological Memo. No. 73, 1975.
3. G. Manley, 'Average frequency of snow-cover in relation to the mean temperature of winter months; based on nine upland stations', *Climate and the British Scene*, Fontana, 1962.
4. G. Manley, 'Central England temperatures; monthly means 1659 to 1973', Q.J. 100, pp. 389–405 and W. 1974 onwards.
5. M.O.
6. Mrs Swaffield, Proudman Oceanographic Laboratory, Bidston.
7. M.W.R. (Annual Summary). Extremes from 1871 for Kew and Falmouth; from 1880 for Braemar; from 1911 for Cambridge and from 1936 for Edinburgh, Royal Botanic Garden.
8. M.O. . J.D.C. Webb and G.T. Meadon, 'Britain's highest temperatures by county and by month,' W. 48, pp. 282–91.
9. M.W.R. and J.M.
10. M.W.R. Table VIII.

12 | Jack Frost

Frost called to water 'halt'.

We accept the fact that the air becomes colder as we climb hills and mountains: snow on the peaks in warm climates, indicative of freezing conditions, is a phenomenon well known to the ancients, too. But there are some very vital aspects of the behaviour of cold air which were not understood two centuries ago and even today are not all that widely appreciated.

One of the earliest investigations into frost was by the naturalist Gilbert White who wrote in his book *The Natural History of Selborne* as follows concerning the intense frost of December 1784: 'On the 10th, at eleven at night, though the air was perfectly still, *Dollond's* glass went down [at Selborne, in Hampshire, not far from Alton] to *one degree below zero* [Fahrenheit or −18°C]. This strange severity of the weather made me very desirous to know what degree of cold there might be in such an exalted and near situation as *Newton*. We had, therefore, on the morning of the 10th, written to Mr.—, and entreated him to hang out his thermometer, made by *Adams*; and to pay some attention to it morning and evening; expecting wonderful phænomena, in so elevated a region, at two hundred feet or more above my house. But, behold! on the 10th, at eleven at night, it was down only to 17 [17°F or −8°C], and the next morning at 22 [22°F or −6°C], when mine was at 10 [10°F or −12°C]! We were so disturbed at this unexpected reverse of comparative local cold, that we sent one of my glasses up, thinking that of Mr.— must, some how, be wrongly constructed. But, when the instruments came to be confronted, they went exactly together: so that, for one night at least, the cold at *Newton* was 18 degrees [18°F or 10°C] less than at *Selborne*; and, through the whole frost, 10 or 12 degrees [Fahrenheit, or 6° or 7°C] [less]; and indeed, when we came to observe consequences, we could readily credit this; for all my laurustines, bays, ilexes, arbutuses, cypresses, and even my *Portugal laurels,* and (which occasions more regret) my fine sloping laurel-hedge, were scorched up; while at *Newton*, the same trees have lost not a leaf!' [1]

Apparently this was the only reported occasion when Gilbert White investigated the temperature at night in any detail. If he had made a practice of taking the night-time thermometer readings he would have discovered that on a calm clear night the valley is usually chillier than the hilltop at any season of the year. In other words, a temperature inversion is said to exist. In 1814, Dr William Wells, FRS, gave the correct solution to the problem. Cold air, being heavier, sinks and collects in the hollows: so that a valley bottom may well be cooler than the free air up to, say, 800 metres (2600 feet). Well over a century later, in the autumn of 1935, at a meeting of fruit growers at the East Malling Research Station in Kent, it became obvious that very little was known about some aspects of frosts in Britain. In 1895 a severe cold wave affected the citrus groves in Florida. They soon discovered that the worst damage was in the hollows so they learnt to avoid them; but no one of importance thought that simple discovery was relevant to Britain. The disasters of May 1935, when killing frost was widespread, had taken growers by surprise. It was not, apparently, widely appreciated why some orchards (those on the higher slopes) had escaped serious damage. Now, nearly two hundred years after Gilbert White's discovery that his thermometers were accurate after all, the successful market gardener and fruit grower is well aware of the danger of hollows (Fig. 72). These are the places where a downhill or katabatic flow of cold air comes to rest.[2]

Fig. 72. **Frost in the valley.** In windy weather, the temperature will usually fall with height, so that the hilltops will be colder than valleys. In still conditions, the reverse can occur. Only the shaded part is liable to experience frost in early autumn and late spring, when the night sky is clear and the wind is light. These circumstances are favourable for radiative cooling.

Fig. 73. **The local nature of frost.** Frost can occur on some surfaces and not on others.

Air frost is said to occur when the temperature in the screen falls below 0°C. Ground frost, like air frost frequently a phenomenon of calm clear conditions, is said to occur when frost is recorded at ground level but the air at screen level is still above freezing. It will of course also occur when the screen temperature is below 0°C.

Unfortunately, the term 'ground frost' is not a very good one (though the expression is in general use and it is difficult to think of anything better) because it frequently happens that objects way above the ground are covered in white when there is no air frost (Fig. 73). You could say, 'There is frost on the roof of the garage and that is 2 metres high, so there must be an air frost.' But like other things connected with weather, it is not quite so straightforward. Here we can bring in the word 'radiation' to help us. Some objects, such as the windscreen and roof of a parked car, the roof of a house, dry sand, long grass, are good conductors of heat, which means that they radiate away a proportion of their heat freely, thus causing their temperature to fall, in turn chilling air in contact with them. What is often not appreciated is that, though the garage roof may be covered in white frost, the free air at the same height only a few metres away may well be above freezing. In the evening, frost will form first on the ground and on surfaces which lose heat most rapidly. But as the night passes, the depth of frosty air will increase as successive layers of air are chilled by mixing and conduction: eventually there may be both a ground (or radiation) frost and an air frost. In a really cold spell, the air will be below freezing from the ground right up to the stratosphere. However, in spring and autumn, there will be a zone where there is no frost between the cold of the mountain tops and the cold of the valleys. Sometimes frost occurs on the hilltops but not the valley bottoms. This is most likely to come about when cold air sweeps in over a warm ground surface, the warmth of the ground keeping the lowest layers of air above freezing.

Good conductors of heat cool down and heat up more rapidly than less efficient conductors such as damp ground or stone walls. So it is to be expected that on some occasions these surfaces will be out of phase with the air around. In 1963, the stone wall in the author's garden was covered in thick hoar frost for several days after the thaw had removed frost from all other surfaces.

Over the ten years 1963–72, Gorleston, Norfolk, had an annual average of sixty-one ground frosts compared to thirty-one air or screen frosts. So on average there might have been frost on the garage roof on thirty mornings when there was no air frost. At Oxford a similar situation may occur on as many as fifty mornings. The proportion of air frosts to ground frosts seems to be around one to two in many places, though in St. James's Park, London, and at Ilfracombe there are nearly three times as many ground frosts as air frosts. At Guernsey, the proportion is over one to four, eight to thirty-six, while at Buxton there are seventy-one air frosts to ninety-one on the ground. It seems probable that open sites have the biggest difference between grass and screen, while more enclosed rural places tend to have a higher proportion of air frosts to ground frosts. In the Highlands, Braemar had 190 ground frosts in 1962. More than one night in two, therefore, was frosty at the level of growing plants. A recent innovation is the recording of minimum temperatures over concrete, the thermometer lying just above the surface. An average of four recent years at Kew gave seventy occasions below 0°C over concrete, but 111 over grass. Clearly the concrete retains its heat more readily since it has high thermal capacity. This helps to explain why urban areas are less frosty.

White frost (sometimes known as white dew) is usual when the dew deposited earlier in the night freezes, or when condensation occurs below freezing point near the
ground. When the white frost is fairly thick, people tend to call it hoar frost. If frost and
fog occur together, hoar frost will be widespread. Sometimes rime will be deposited as
well, when supercooled water droplets touch objects below freezing point and freeze
immediately. Such deposits can be very beautiful if the sun shines. Notably
widespread rime occurred over Christmas 1944, in early December 1962, in Decembers
1976 and 1981 and in early January 1993. So thick were the deposits of rime in 1962
that a breeze which cleared the fog left small drifts of ice crystals, looking just like
snow. At first sight it seemed as if snow had fallen and collected only under the hedge.

From 28 to 30 January 1976, Polar Continental air, bringing a 'black frost', covered
much of southern England. Many inland places had temperatures below freezing by
day and night so that the ground was frozen hard, though no sign of frost was apparent. The landscape was still green. Life appeared to be stirring in the trees and the sun
shone brightly. Many children and quite a few adults were not prepared to believe that
the temperature was well below freezing, as the usual whiteness was not to be seen.
Motorists, who had to leave their cars parked outside overnight, heard the frost warnings and took precautions, covering their windows and windscreen with paper. In the
morning the glass was clear and the paper, if it had not blown away, was still dry. Even
on the way to work virtually no steam formed on the inside of the car windows. It was
a classic example of a black frost which is most likely to occur with very dry continental air that has taken a short sea track, particularly in the south of England. The term
'black' is used to distinguish this sort of frost from the whiteness usually associated
with freezing conditions. Sometimes people talk of a 'wind frost', but this is not necessarily the same thing as a black frost because the wind may be carrying moist air. The
most important point about a wind-borne freeze-up is the penetrating power of the
frost. A 50 k.p.h. (31 m.p.h.) wind at a temperature of –1°C (30°F) has the same freezing power as still air at –3°C (27°F). The same wind speed and an air temperature of
–4°C (25°F) (like 1 February 1956) is as penetrating as still air at –16°C (3°F). This indicates the 'wind chill factor'.[3] It is the wind which brings about the extra cooling.
Notable visitations of 'black frost' occurred in February 1956 and in 1947 and 1963,
though lying snow gave visible evidence of the cold in many parts. Glazed frost or
freezing rain is also uncommon and will be described in Chapter 17. It is rain which
falls onto a surface which is below freezing point. Black ice is a popular alternative
word for glaze, often used with reference to its occurrence on road surfaces.

As a rule, frost is accompanied by calm cloud-free weather, although on the top of
the Pennines, for example, cloud, wind and freezing conditions are frequent.
Warning signs for frost during the night are: air of polar origin, little or no wind, a
clear sky and low humidity before sunset. The humidity of the air is important, for
the drier the air the more rapid the night-time cooling. If the air is damp, fog may
form before the temperature drops to freezing point. Such fogs may prevent further
cooling, or even cause a rise in temperature. Quite frequently a clear evening will
give way to cloud before morning. As the clouds approach, the temperature will rise,

so that it may be warmer at sunrise than dusk. Sometimes the thermometer does not stay below freezing point continuously, rising above 0°C when a cloud moves across and falling again when the cloud passes by.

Sandy regions are especially prone to frost for the ground tends to be dry. At Santon Downham, where the weather station occupies a sandy hollow in Breckland on the Norfolk-Suffolk border, an air temperature of −7·7°C (18°F) has been recorded in May (Table 24). This would be distinctly newsworthy if it happened in London in midwinter. Places affected by frequent frost in contrast to other parts nearby are called frost hollows or frost pockets. Such hollows are numerous. Whether they are notorious for severe frosts, or have frost a little earlier in the autumn and a little later in the spring, will depend on the configuration of the land and the nature of the soil. A long narrow valley draining a large area of high ground will be much more frost prone than a wide shallow one, for the severity of the frost often depends on the size of the donor area, i.e. of the ground which gives its cold air to the land below.[4]

The incidence of air frost during one particular May over south-east England is shown in Fig. 74. The coast, central London and the hilltops were all frost-free. Except in unusually severe cold snaps they are likely to remain so in every May.[5]

Instrumental recordings exist for a number of cold spots, of which the Rickmansworth frost hollow, a long narrow valley in the Chilterns 29 kilometres (18 miles) north-west of London, was among the first to be investigated. It is a pity that the record was only kept from 1938 to 1950, but this was long enough to produce some startling facts. Air frost occurred in three Augusts and nine Septembers, or more

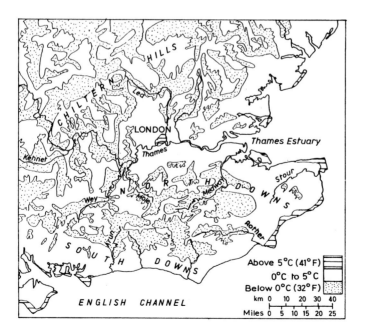

Fig. 74.
Minimum temperatures experienced during a frosty May

than one September in two. In April 1938 there were twenty-one nights of air frost. May 1941 had thirteen. In the thirteen years no May was frost-free, while Kew has had only eight Mays with frost in seventy-five years. The average annual number of air frosts at the site is 124, fourteen more than the small Highland town of Braemar.

During the period of observations, there were few houses in the valley. Since then much of it has been built upon. It seems highly probable that the presence of the houses could affect the flow of cold air down the valley and so considerably lessen the risk of frost. Between July 1961 and May 1979, records were kept in the valley and no air frosts were recorded in any of the summer months, and ground frosts only in two years, in June 1962 and 1975. It does seem that greater urbanisation could have effected some change. It appears likely that the existence of buildings in the valley would drastically alter the free flow of cold dense air, often only a matter of a few metres deep. There are some plans afoot to keep weather records again, if a suitable site can be found. It would be interesting to see how the results compare with the long-standing records nearby and with the earlier data. Information collected at Santon Downham since 1958 suggests that that place may well be more frosty than Rickmansworth (even before Rickmansworth was built up) as summer minima seem significantly lower, while the less extreme values for midwinter suggest a lack of ideal conditions for sensational records. Generally the 1962–63 winter did not give such widespread low minima as either 1940 or 1947. In the north of England there is a notable frost hollow at Houghall in the Wear valley near Durham, where −21°C (−6°F) has been recorded in March. Braemar, in the valley of the Upper Dee in eastern Scotland, is also noted for low records. Blackadder in the Tweed valley has recorded −30°C (−22°F), but the exposure was not standard. The thermometer was not the official height above the ground (a matter of considerable importance as we have seen), so that the official record goes to Braemar with −27·2°C (−17°F) on 11 February 1895 and also on 10 January 1982. Kelso has recorded −26·7°C (−16°F) on 17 January and −24·4°C (−12°F) on 18 January 1881.

Dry, clear calm conditions with air of polar origin are ideal for frost formation. However, the truly sensational values are only likely if the ground is snow covered as well. Just how intense is a temperature of around −20°C (−4°F)? The author's experience with a recording of −18°C (0°F) will give us a standard. On 21 February 1948, Polar Continental air had brought a short but severe cold spell in an otherwise mild winter (see also page 195). Just after sunset the snow stopped, the clouds broke up, the air fell still and the stars shone very brightly, indicative of very dry clear air. Radiative cooling in the already cold air was intense. By 21 h, the screen temperature had fallen to −13°C (9°F) and a thermometer lying on 10 cm (4 in) of powder snow registered −18°C (0°F). Stories about water freezing virtually instantaneously at 0°F have always aroused the author's curiosity. To be able to put this to the test in Sussex was an unexpected opportunity. A one-pound jam jar was filled with water from a tap indoors and placed on the ground, standing level in the snow. The water in that jar froze so rapidly that the surface of the ice was inclined at an angle of about 25° to the horizontal, numerous air bubbles being imprisoned inside the ice. In other words freezing was so rapid that the water froze before it had anywhere near stopped swinging about in the jar. To prevent it freezing on its way from the house to the garden (a distance of 20 metres at the most), the jar was carried under a coat. When the thaw came a few days later, a small pile of powdered glass remained on the grass as a memorial to this sensational piece of freezing, sensational to us British, though commonplace in our latitudes in Canada and the northern states of the USA. The events described took place at Hurstpierpoint, about 16 kilometres (10 miles) north of Brighton in the Sussex Weald.

Pilot surveys in the Chipstead valley,[6] 24 kilometres (15 miles) from central London, revealed night temperatures on a par with the Scottish Highlands and with Rickmansworth before it was urbanised. Ian Currie wrote on 7 January 1985 that 'a ridge was moving south over the area in a very cold continental air mass and there was 8 cm (3 in) of snowcover. I made my way to a suitable site in a field adjacent to some houses. It was 11 p.m. and the night was still. The snow lay crisp and frost covered. Clouds of vapour rose from manhole covers along a nearby road. Using a pocket whirling hygrometer, a dry bulb reading gave −17°C (1°F) and I mused what a permanent weather station would have read the next morning, although low cloud did spread across later in the night.'

Two further examples confirmed my suspicions that this was a very interesting spot. At 05.55 on 30 October 1988 a reading of −4·5°C (23·9°F) was obtained, compared with 0°C at Gatwick and of −2·2°C (28°F) near the summit of Reigate Hill, 230 metres (755 feet) high and 5 kilometres (3 miles) to the south. On 5 June 1989 the valley was etched in white with a reading of −1°C (30°F), a figure some inner London weather stations did not attain in January that year.

The dry valley runs north-east from the crest of the North Downs. In October 1990 a permanent weather station was set up. November 1990 had fourteen air frosts in the valley. There were two at Coulsdon nearby. In 1992 the valley experienced forty-seven air frosts, the south London suburb of Morden had eighteen, Birmingham University twenty-seven and Aviemore in the Highlands forty-nine.

The last column of Table 25 shows the number of air frosts to be expected in an average year. The largest totals occur in upland valley sites such as Braemar. For most inland parts away from the west coast, the total will be in the range fifty to sixty, with higher totals in the hollows. Along the east coast, frost will occur on between twenty and thirty-five nights, while exposed places on the other side of the country as far north as Tiree will have twenty frosts or less. In Shetland, Lerwick averages forty-nine.

So much for the total number of air frosts in winter. But which month has the most frost? (See Table 25.) January is more likely to be frosty than any of the other months. Rather over a quarter of January nights register temperatures below freezing at Kew. In the open country the proportion rises to about half. At Cardiff and Scilly, February is more likely to be frosty, partly because the influence of the sea on cold air is less then, for the sea is colder. As we have seen, a reading below 0°C can occur locally in June, July and August, but in general places which are reasonably open will be free from frost from the third week of April to mid-October. At Rugby the average frost-free period lasts from 28 April to 22 October. At 700 metres (2300 feet) in the

Pennines, frosts are to be expected in May and September. April in the Pennines is notorious for cold nights (in contrast to the lowlands) for the upper air has not warmed up appreciably.

Averages also hide interesting facts: they do not tell us what happens in very cold or very mild winters, for example. A further set of statistics shows the number of frosty nights each year for a number of places (Table 26). The range shown is most surprising: it varies from 3 to 32 at Penzance; 3 to 61 at Colwyn Bay; 8 to 73 at Kew; 28 to 100 at Goudhurst; 12 to 90 at Lowestoft; 17 to 83 at Edinburgh, and 8 to 80 at Stornoway. At Scilly (not shown in the updated table because the weather station closed), the famous winter of 1962–63 gave only twelve frosty nights, a total exceeded in an average January at Wisley, Surrey. Even more remarkable is the fact that nine winters out of twenty were free of air frost. No wonder the Scillies and west Cornwall are able to supply London with spring flowers and vegetables in midwinter. It seems fantastic, but Kew actually had one more frosty day (an 'ice day', maximum < 0°C) than Scilly had frosty nights in the winter of 1962–63.

At Kew, the low total of eight in 1960–61 sent the author back to recheck the statistics. February of that winter had no frost at all. As expected, the 1962–63 winter had many frosty nights, though a study of the figures showed that it did not have the most, the honour going to the fairly unremarkable winter of 1969–70 which had one more, but was not so severe. Goudhurst, typical of a low-lying rural station, is very frosty by British standards, having more night frosts than Edinburgh. Lowestoft on the east coast is markedly less frosty than Cambridge and often less frosty than Ringway. One may find this hard to accept, for the east coast is well known for being bitterly cold in winter. So it is, but wind and frost do not often go together, though the wind chill factor may be higher at Lowestoft. Ringway airport, Manchester, is usually less frosty than Cambridge. This would be logical because of its more westerly situation, open to mild Atlantic air-masses, and less liable to be affected by Polar Continental air. From these figures, it seems that in a mild winter the number of frosts to be expected is about twenty in inland places which are reasonably level. In undulating country, forty seems more likely in the valleys and less than ten in Inner London.

So far we have generally assumed that frost is only a night and early morning event. Occasionally the temperature can remain below freezing all day. Table 27[10] shows the average number of days when this happens, often referred to as 'ice days'.

The average over the lowlands of the east is between two and four a winter, in the south-west as few as one day in two years. The higher average and maximum total at Hastings, compared to Gorleston and Acklington (in Northumberland), show the ameliorating effect of the wider sea on cold winds. Further north still, at Craibstone near Aberdeen, the average number of ice days increases fractionally, probably a result of being nearer Arctic airstreams. In spite of this, the maximum number of ice days recorded near Aberdeen in a winter is under half the total of Hastings. At all the places shown, several winters had no ice days at all. In the thirty-year period there were three winters without ice days at Craibstone, twelve at Hastings and Kew and over twenty at Falmouth and Valley (Anglesey). Birmingham, on the other hand, has

Table 24 **Monthly Extremes of Temperature for Two Frost Hollows**

Temperature (°C)

RICKMANSWORTH 1938–50

	Jan	*Feb*	*Mar*	*Apr*	*May*	*Jun*	*Jul*	*Aug*	*Sep*	*Oct*	*Nov*	*Dec*	*Year*
Highest recorded	14·4	18·3	22·8	30·1	33·3	34·4	35·0	34·4	32·2	25·6	21·7	15·0	35·0
Lowest recorded	−20·0	−19·4	−17·2	−8·9	−8·3	−2·8	0·0	−2·2	−3·9	−9·4	−10·6	−18·9	−20·0

SANTON DOWNHAM 1958–87

	Jan	*Feb*	*Mar*	*Apr*	*May*	*Jun*	*Jul*	*Aug*	*Sep*	*Oct*	*Nov*	*Dec*	*Year*
Highest recorded	14·5	18·3	25·0	25·6	28·3	30·0	33·3	33·5	29·2	28·0	17·8	15·8	33·5
Lowest recorded	−18·9	−15·0	−12·2	−10·6	−7·7	−5·6	−1·1	−1·7	−5·6	−7·5	−12·1	−16·1	−18·9

Because of their sheltered sites frost hollows are extreme places, but not only at night, as the figures show. Santon Downham holds the March absolute maximum for the British Isles.

Table 25 The Monthly Distribution of Air Frost

Height above sea-level (m)

Average number of days of frost

		Jan	Feb	Mar	Apr	May	Jun	Jul	Aug	Sep	Oct	Nov	Dec	Year
	SCOTLAND													
82	Lerwick	9·1	10·1	8·9	4·6	0·6					1·0	6·4	8·0	48·7
3	Stornoway	8·6	10·0	8·6	5·3	0·9				0·2	1·1	7·8	8·3	50·8
9	Tiree	4·9	6·0	2·7	1·8						0·1	1·0	2·9	19·4
4	Inverness	11·6	13·9	6·1	3·0	0·1				0·4	0·5	7·1	9·0	51·3
8	Fort William	11·4	13·1	9·4	5·0	0·7				1·5	1·9	8·9	11·7	62·5
339	Braemar	18·9	19·7	17·9	10·2	3·5	0·3		0·7	1·5	5·2	14·5	15·7	108·1
52	Aberdeen	10·9	13·4	8·2	5·0	0·3					0·9	8·6	10·8	58·1
23	Edinburgh (RBG)	11·6	12·1	7·2	3·9	0·5				0·1	1·0	8·7	11·1	56·2
107	Glasgow	12·4	13·6	8·9	3·3	0·4					0·6	9·1	10·1	58·4
242	Eskdalemuir	15·8	17·5	14·8	9·0	2·9			0·4	0·8	3·1	12·5	13·7	90·5
	ENGLAND													
29	Tynemouth	6·1	7·4	3·4	1·7	0·1						2·3	5·6	26·5
20	York	12·4	12·4	5·9	2·9	0·1					0·7	7·2	11·2	52·8
25	Lowestoft	12·1	9·3	5·7	1·9							3·0	7·7	39·8
12	Cambridge	12·6	10·7	7·8	3·0	0·2	0·1			0·1	0·5	5·9	10·3	51·1
5	London (Kew)	9·7	8·0	6·1	2·3	0·4				0·1	0·7	5·1	8·3	40·7
4	London (St. James's Pk)	8·0	6·1	3·6	0·7							2·8	6·6	27·8
85	Goudhurst	14·8	13·9	9·0	6·0	0·8				0·1	1·4	7·9	13·2	67·1
8	Worthing	9·6	7·6	4·3	0·9							2·5	6·7	31·6
5	Ryde, Isle of Wight	8·0	6·0	3·3	0·4						1·0	5·3		24·0
126	Salisbury	11·9	13·0	8·3	3·7	0·3					0·7	6·5	11·0	55·4
32	Exeter	9·6	9·2	5·6	2·2	0·5					0·4	5·5	8·4	41·4
36	Plymouth	7·0	6·9	3·7	0·8	0·1						1·7	5·5	25·7
51	Falmouth	4·7	4·1	2·4	0·3							0·7	2·9	15·1
8	Ilfracombe	5·2	3·6	1·1	0·1								1·2	11·2
48	Bristol (Long Ashton)	10·1	10·9	5·7	2·5	0·2					0·4	4·7	9·0	43·5
65	Cheltenham	10·3	10·3	6·4	2·4	0·1					0·4	5·2	9·1	44·2
163	Birmingham (Edgbaston)	10·5	10·7	6·7	1·5	0·1						2·7	7·4	39·6
76	Manchester (Ringway)	10·5	11·1	6·1	2·4	0·2					0·5	5·1	8·6	44·5
556	Moor House	21·2	23·0	21·2	14·8	5·7	0·7	0·5	0·3	2·0	3·9	16·4	20·0	129·7
7	Morecambe	8·7	9·3	3·7	0·8							3·7	6·4	32·6
26	Carlisle	11·0	12·1	8·3	5·4	0·5				0·3	0·9	9·4	10·9	58·8
	WALES													
67	Cardiff (Rhoose)	9·4	10·4	6·4	2·4	0·2					0·1	4·1	7·8	40·8
8	Swansea	6·7	6·7	3·8	0·7							1·7	4·5	24·1
133	Aberporth	7·4	7·1	2·8	0·6							1·7	4·6	24·2
36	Colwyn Bay	7·6	7·7	3·1	1·1							1·2	4·0	24·7
	ISLE OF MAN													
87	Douglas	5·1	6·6	2·3	1·1							1·5	3·8	20·4
	CHANNEL ISLES													
83	Jersey Airport	5·0	4·5	2·0	0·1							0·1	3·5	15·2

Table 26 Number of Air Frosts from 1959–60 to 1992–93 [8]

	Stornoway	Aberdeen	Glasgow	Edinburgh	Eskdalemuir	Durham	Ringway	Edgbaston[9]
1959–60	41	67	48	32	72	53	36	25
1960–61	35	64	47	31	70	44	24	13
1961–62	78	85	66	58	98	92	55	54
1962–63	48	88	82	79	126	99	87	82
1963–64	37	53	44	33	67	62	39	37
1964–65	50	85	65	45	108	80	47	41
1965–66	57	88	62	58	93	79	43	38
1966–67	40	61	58	30	63	39	26	17
1967–68	61	91	75	58	101	82	59	44
1968–69	59	65	69	54	102	67	50	53
1969–70	80	96	84	67	107	75	58	60
1970–71	34	46	42	22	75	42	31	24
1971–72	42	48	46	32	64	40	20	18
1972–73	46	79	58	39	90	48	33	21
1973–74	32	68	48	31	85	46	19	22
1974–75	29	65	50	32	87	45	22	18
1975–76	25	49	48	33	76	40	29	34
1976–77	41	79	66	36	96	61	36	21
1977–78	50	66	68	52	84	53	36	35
1978–79	56	80	86	83	112	78	72	79
1979–80	25	55	69	39	98	41	36	36
1980–81	37	60	65	54	82	57	36	37
1981–82	41	69	64	49	96	63	40	50
1982–83	41	65	69	46	82	50	34	41
1983–84	27	62	64	49	104	62	33	44
1984–85	30	59	63	47	78	65	49	53
1985–86	43	75	85	76	93	76	59	58
1986–87	34	61	68	48	110	48	48	44
1987–88	29	55	58	30	74	41	23	26
1988–89	21	34	44	26	79	31	37	30
1989–90	16	42	40	17	60	39	31	13
1990–91	46	56	71	42	85	53	57	43
1991–92	8	56	54	41	72	52	32	30
1992–93	21	46	57		93	56	38	34

Table 26   **Number of Air Frosts from 1959–60 to 1992–93** continued

	Cambridge	Lowestoft	Kew	Goudhurst	Penzance	Rhoose	Colwyn Bay	Aldergrove	Jersey
1959–60	53	24	24	64	9	26	13	51	9
1960–61	41	17	8	50	13	21	11	38	2
1961–62	83	50	41	85	15	54	42	64	21
1962–63	97	90	72	100	32	91	61	70	47
1963–64	56	42	24	58	18	38	21	36	20
1964–65	61	50	31	85	20	48	34	62	17
1965–66	47	39	24	60	7	35	24	47	10
1966–67	41	27	17	49	7	25	8	29	7
1967–68	65	41	32	81	16	48	30	59	7
1968–69	47	36	48	74	19	43	28	53	15
1969–70	58	52	73	79	9	57	30	71	15
1970–71	42	26	49	57	17	24	15	33	16
1971–72	22	16	43	47	9	17	6	25	3
1972–73	52	26	53	68	9	31	9	44	8
1973–74	44	25	47	43	6	22	3	36	5
1974–75	27	16	29	40	3	17	3	35	6
1975–76	48	33	34	66	8	31	14	28	10
1976–77	56	39	50	60	16	31	17	52	7
1977–78	48	29	41	48	6	32	23	47	14
1978–79	56	61	64	77	17	60	42	69	21
1979–80	25	21	52	55	15	29	21	43	10
1980–81	47	38	49	67	11	28	16	33	8
1981–82	65	39	46	63	4	34	35	36	9
1982–83	50	29	39	54	10	36	27	50	17
1983–84	55	21	50	56	13	37	16	46	1
1984–85	65	49	62	60	17	48	36	63	23
1985–86	78	55	63	78	22	51	28	64	27
1986–87	60	50	53	61	18	46	29	44	30
1987–88	30	14	32	37	1	21	9	23	5
1988–89	29	13	36	32	9	18	7	27	0
1989–90	19	12	22	28	2	16	18	24	5
1990–91	44	22	39	47	12	45	27	45	20
1991–92	44	28	51	47	11	34	18	27	9
1992–93	36	17	39	41	8	23	12	31	9

the dubious distinction of having some ice days in twenty-eight out of thirty winters.

One may well have reason to doubt some of the figures for ice days. Surely there are more than three complete days a winter at Kew, for example, or more than seven in Birmingham, when ice does not melt and the grass is white all over from dawn till dusk? Some explanation is perhaps needed, as the figures do not agree with the casual observations of the layman. On 10 and 11 December 1976 at Guildford, the ground remained frozen all day, yet the maxima were 1° and 2°C (34° and 36°F) respectively.

The definition of an ice day ought to be easy. But to the layman it is also an 'ice day' when the ground remains white all day, even if the thermometer says something different. Curiously, what is frozen will remain frozen, but what is not frozen will not freeze until the temperature goes below freezing again. These are not, of course, true ice days, as defined by the Meteorological Office. So far as can be ascertained no one has collected any information on these pseudo-ice days.

Table 27 shows that the number of ice days has declined comparing 1941–70 with 1971–90. The number is halved at Southampton, down by two-thirds at Hastings and more than halved at Lerwick, for example. These figures suggest less severe winters, possibly due to global warming; but it is not statistically fair to compare a thirty-year period with twenty years. Between 1991 and 1995 the number of ice days has not redressed the balance. February 1991 notched four or five in the south and November 1993 one or two, plus others in fog. December 1995 had one or two ice days in the south.

Table 27 Average Number of Days Below 0°C All Day (Ice Days) for the Periods 1941–70 and 1971–90

	Average per winter (1941–70)	Most in a winter (1941–70)	Average per winter (1971–90)	Most in a winter (1971–90)
Lerwick	5·0	19	2·1	7
Stornoway	1·0	6	0·2	2
Aberdeen, Craibstone	3·6	12	2·8	13
Rothesay	0·7	4	0·16	4
Abbotsinch	3·6	12	2·7	9
Eskdalemuir	9·5	36	5·8	21
Douglas, Isle of Man	0·6	8	0·5	3
Acklington	3·1	19	1·1	4
Gorleston	3·2	23	1·5	12
Cambridge	3·2	30	1·9	9
Birmingham*	7·3	32	4·8	17
Kew	3·3	21	2·0	11
Hastings	4·1	26	1·3	11
Southampton	2·0	13	1·0	7
Valley	0·3	4	0·4	3
Falmouth	0·6	6	0·5	4
Aldergrove	1·2	7	0·9	3

* Edgbaston to 1979, Elmdon from 1980

NOTES TO CHAPTER 12

1. Gilbert White, *The Natural History of Selborne*, Thames & Hudson, 1993. E.L. Hawke, 'Frost hollows', W. 1, pp. 41–5.
2. R. Bush, *Frost and the Fruitgrower*, Cassell, 1947, especially Chapters VII to XI.
3. H.L. Penman, 'Effect of wind and temperature on freezing power', W. 2, p. 141.
4. R. Bush, 'Classification of frosty areas', ibid., Chapter IX.
5. The author, 'Frosts in May', W. 7, p. 40.
6. I.M. Currie, 'A Remarkable Surrey Valley: Its Reputation for Extremes Confirmed', J.M. 17, pp. 220–3.
7. M.O. 3.
8. M.W.R.
9. John Kings, Birmingham Climate and Atmospheric Research Centre (BCAR) - Weather Service (Manager), School of Geography, University of Birmingham. A.H. Weeks, 'Frost Hollows', J.M. 17, p. 344.
10. M.W.R.

13 | Sensational Months and Seasons

Occasionally we experience months of abnormal warmth or cold. In 1916, 1917 and 1919, March was colder than an average January. In 1938, March was warmer than an average April. The Januarys in 1916, 1921, 1974 and 1975 were nearly as warm as an average April in many places. October 1974 was colder than the following December. November 1994 was frost-free, even of ground frost, over wide areas.

Winter Cold

Since 1795 there have been twenty-three months in central England [1] with a mean temperature below 0°C. These comprised five Decembers, eleven Januarys and seven Februarys. The coldest January had a mean temperature of −3·1°C (26·4°F) in 1795. February was coldest in 1947 with a mean of −1·9°C (28·6°F); and December in 1890 with −0·8°C (30·6°F). There were twelve sub-zero months in the nineteenth century. In the twentieth century (to 1994) there have been eight: February 1986; December 1981; January 1979; January and February 1963; February 1956; February 1947 and January 1940. Means for the period 1931–60 are: January 3·4°C (38·1°F), February 3·9°C (39°F), December 4·7°C (40·5°F), giving a range below average of 6·5°C (11·7°F), 5·8°C (10·4°F) and 5·5°C (9·9°F) respectively.

February 1895 had a mean temperature of −1·6°C (29·1°F) at Kew, −4·1°C (24·6°F) at Buxton, and −5·8°C (21·6°F) at Braemar, or 6·4°C (11·5°F) below average. On 11

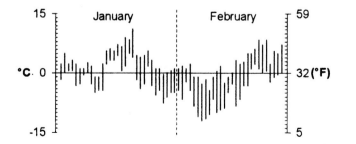

Fig. 75.
The daily maximum and minimum temperatures in the severe winter of 1895 (Brixton, London)

Fig. 76. **Skating on the River Dee at Chester in February 1929**

February, Braemar reported −27·2°C (−17°F), which is still the lowest official screen reading in the British Isles. (It was equalled in January 1982.) There were ten nights below −17·8°C (0°F) in a sequence of fourteen. On Lake Windermere the ice was reported to be 60 cm (2 feet) thick. In London skating lasted seven weeks. The cold was so intense that much damage was done to water mains and new bye-laws were introduced so that thereafter mains were sunk deeper. Fig. 75 shows the daily maximum and minimum in London during this severe winter. The short mild spell, the greater range of temperature and the colder nights compared to 1947 and 1963 should be noted.

February 1929 was very cold and the temperature fell to −18°C (−0°F) in Herefordshire, though some mild days at the end of the month raised the average sufficiently to keep it out of the list of sub-zero months given earlier. The ice on Lake Windermere was reputed to have borne the weight of 50,000 skaters. Conditions were ideal for good skating as there was little snow-cover to spoil the surface. Fig. 76 shows people skating on the Dee in Chester. December 1933 was remarkable for its severe cold in the south. An anticyclone, centred over the north of England, produced a flow of cold Polar Continental air in the south, while the north had milder

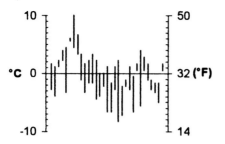

Fig. 77. (*above*) **The daily maximum and minimum for Kew for January 1940**

Fig. 78. (*right*) **Mean temperature °C for January 1940**

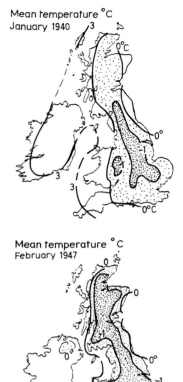

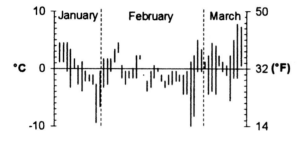

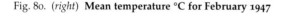

Fig. 79. (*above*) **The daily maximum and minimum at Kew in the cold winter of 1947**

Fig. 80. (*right*) **Mean temperature °C for February 1947**

Tropical Maritime air from the Atlantic. Not surprisingly, the warmest places in Britain were the Hebrides and Shetland with mean temperatures of 7°C (45°F), compared with only 2°C (36°F) in London and Bournemouth and 1°C (34°F) at Tunbridge Wells. The Thames froze over above Oxford. At Tavistock on the western edge of Dartmoor the thermometer fell to −7°C (19°F) on the coldest night of the month, making a strange contrast with Orkney where no air frost was recorded.

January 1940 was the coldest month for forty-five years in many places (Figs. 77 and 78, and Table 28). Nearly all England and Scotland had a mean below 0°C. There were fourteen ice days at Cranwell, Lincolnshire, and twelve each at Edgbaston and at Lympne near Folkestone, though there was a short mild spell in the month when temperatures exceeded the average.

February 1947 was fractionally colder than February 1895 over central England, but not at Kew or Manchester (Figs. 79 and 80). At Kew, the highest recorded temperature in the month was 5°C (41°F), below the average maximum by some 2·5°C

Table 28 **The Mean Temperature of the Coldest Winter Months at Manchester since 1890** [2]

	°C	°F
Feb 1895	−1·7	28·9
Feb 1947	−1·4	29·4
Jan 1963	−1·4	29·5
Jan 1940	−0·8	30·5
Jan 1945	−0·1	31·9
Feb 1986	−0·1	31·9
Dec 1890	0·0	32·0
Feb 1963	0·0	32·0
Feb 1942	0·2	32·3
Jan 1979	0·3	32·5
Feb 1956	0·4	32·7
Dec 1950	0·5	32·9
Dec 1981	0·7	33·3

The Coldest Februarys at Kew (date order) [3]

	Average maximum		Average minimum		Mean		Days below 0°C all day
	°C	°F	°C	°F	°C	°F	
1986	2·0	35·6	−2·2	28·0	−0·2	31·6	4
1963	2·2	36·0	−1·4	29·5	0·4	32·7	12
1956	2·6	36·7	−2·7	27·1	−0·1	31·9	11
1947	0·5	32·9	−2·7	27·1	−0·9	30·3	13
1942	2·7	36·9	−1·6	29·1	0·5	32·9	5
1929	3·4	38·1	−2·5	27·5	0·3	32·5	7
1895	1·4	34·5	−4·5	23·8	−1·6	29·2	12
1886	3·1	37·5	−0·8	30·5	1·1	33·9	2
Average 1941–70	6·8	44·1	2·3	36·1	4·5	40·1	

(4·5°F). The average maximum for the month was just above freezing. There were also 234 hours of continuous frost, almost ten days. At Oxford, frost set in at 18 h on the 10th and continued without a break until 6 h on the 26th. Only in the south-west did the mildest day of the month exceed the average maximum. Falmouth reported 11°C (52°F) in contrast to only 3°C (37°F) at Gorleston on the east coast, where sea temperatures were not far above 0°C. At Kew, the sun shone on only five days to give a total of seventeen hours for the month, compared to the average of sixty-one. The period from the 2nd to the 22nd was completely sunless.

The absence of clear skies prevented radiative cooling; the low night minima were confined to a few days around the end of the month. The combination of sombre grey skies, low temperatures and substantial snowfall made the month a grim one in the period of post-war austerity in Britain. Transport of the already barely adequate supply of fuel was seriously interrupted. Coal stocks at power stations dwindled, as colliers from the north-east coast were delayed by gales, or freezing conditions, at the loading and unloading points. The majority of households still had coal fires and electricity generation was almost totally dependent on coal for the power-station boilers. Gas was also made from coal and pressure could not always be maintained. Widespread restrictions on the use of electricity were introduced, and there were penalties for breaking the regulations. To keep one room warm was the aim of most households.

After a mild January the first day of February 1956 saw a severe blast of air of Siberian origin sweep across Britain, giving day maxima well below freezing in many parts of the south, as low as –6°C (21°F) at Ipswich. Even in the Scillies there were two days of continuous frost and gales. The month was not as cold as February 1947: there was much more sunshine and several mild days. But February 1956 will be remembered by arable farmers for the severe damage to winter wheat, since, except along the east coast, there was very little snow to protect the ground from the hard frost. The remarkable onset of this cold spell on 31 January 1956 is described on page 164.

At Kew and Gorleston, the thermometer failed to rise above freezing point on eleven days. At Hastings and Birmingham, the total was twelve. Even as far west as Falmouth there were six ice days, in contrast to only three at Manchester and one each at Stornoway and Lerwick. Once again the cold in the south was more intense. Among low readings were –17°C (2°F) at Wye, Kent; –15.5°C (4°F) at Bodiam, Sussex; –15°C (5°F) at Shrewsbury and at Stone, Staffordshire. Probably the absence of widespread snow-cover and of still, clear weather at night prevented really sensational recordings. At Heathrow, for example, the ground was only snow-covered on three mornings.

In 1963, there was probably the coldest January since 1814 over England. Fig. 81 shows the very low mean temperatures over the south and the Midlands. The comparative mildness of the west is also noticeable. As in February 1947 the month's highest temperatures failed to reach the average, many places not exceeding 5°C (41°F). At Birmingham there were seventeen ice days, and at Kew thirteen. Falmouth and Holyhead had two and Aldergrove near Belfast only one day of continuous frost. Over most parts there was frost every night, bar one or two, except, for example, Stornoway which had only eleven. The daily maximum and minimum temperatures at Kew for December 1962 and January and February 1963 are shown in Fig. 82. Note the low temperatures in the foggy spell in early December.

The Thames was frozen over above Kingston power station, although not at Tower Bridge because of the warmth from industrial cooling water, which pours into the river. In the estuary there was a kilometre of ice on the foreshore at Herne Bay and ice in Whitstable harbour (Fig. 83). There was ice on the beach in South Wales, at Penarth, near Cardiff (Fig. 84). Unlike in February 1947, there was above average sunshine, and unlike in February 1956, the snow-cover was deep except near some coasts. Outside the towns the snow-cover was continuous until the end of the month,

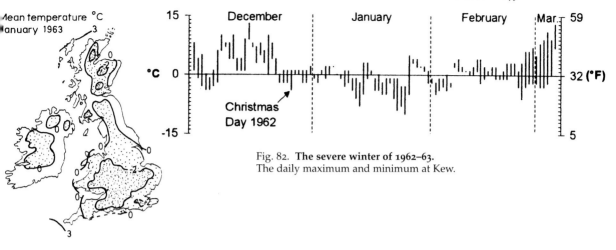

Fig. 82. **The severe winter of 1962–63.**
The daily maximum and minimum at Kew.

Fig. 81. (*above*) **Mean temperature °C in January 1963**

which gives a good indication of the intensity of the cold: some 5°C (9°F) below aver-age. On the south coast a small waterfall which drops to the beach at Kimmeridge, Dorset, was frozen. In February 1963, for the second month in succession, daytime temperatures failed to reach the long-period average in many parts. The frost lasted all day nine times in Birmingham but only twice at Kew.

In many places the winter of 1978–79, which really began in earnest just after Christmas, turned out to be the most severe for sixteen years. The cold was associ-ated with low pressure. Fronts and depressions crossed the country bringing in Atlantic air from time to time so that the cold was not as continuous as in 1962–63. Fig. 86 shows these alternating cold and less cold spells. On the other hand, heavy rain, gales, deep snow, freezing rain and spells of severe frost probably caused more inconvenience than in 1962–63. On the night of 12–13 January 1979, the temperature fell to −24·6°C (−12°F) at Carnwath, at a height of 208 metres (682 feet), near Lanark. It was the fourth lowest temperature on record in the British Isles after −27·2°C (−17°F) at Braemar, 11 February 1895, and −26·7°C (−16°F) at Kelso on 17 January 1881 and 3 December 1879.

South of the Thames conditions were not so severe as in 1962–63 and Fig. 87 shows that the isotherm for freezing did not extend so far south. In Brighton,[4] the four months from mid-November to mid-March had fifty-two nights with frost and eight ice days, compared to seventy-seven and twenty-seven respectively in 1962–63. Over the whole of England and Wales the number of ice days was less than in 1962–63. There were sev-enteen at Birmingham (Edgbaston), seven at Manchester (Ringway), five at Kew and Cardiff, two at Douglas and Valley (Anglesey) and none at all at Falmouth. In Scotland there were nine ice days at Lerwick, eight at Aberdeen and twenty-two at Eskdalemuir. In Northern Ireland there were two ice days at Belfast Airport.

Fig. 83.
**Ice in Whitstable harbour,
Kent, in January 1963**

Fig. 84.
**Snow and ice on the beach
at Penarth, near Cardiff, in 1963**

Fig. 85. (*right*)
A frozen cascade in Cautley Spout waterfall, Cumbria, in 1986

December 1981 was one of the coldest months on record at Glasgow, Abbotsinch, with a mean temperature of −1·7°C (29°F), equalling January 1940 and January 1881. In many inland parts as far south as Hertfordshire, the mean temperature was below freezing. The incredible cold in the Midlands, in particular, with numerous readings below −20°C (−4°F) in January 1982 did not last long and the mild start and end to the month meant that it ended up not far from average (see pages 171–2). The 1985 winter saw two severe spells, one in January and one in February, with much snow.

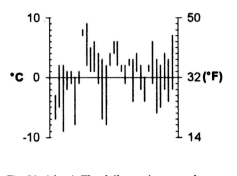

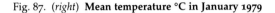

Fig. 86. (*above*) **The daily maximum and minimum at Kew for January 1979**

Fig. 87. (*right*) **Mean temperature °C in January 1979**

For continuous cold, February 1986 has not been held in the general public's memory because of bright sunshine and little snow and inconvenience, except to people who had not lagged their pipes properly. Fig. 85 demonstrates the severity of the cold in Cumbria. The cold was more persistent though not as severe as in February 1956. The month's absolute maximum was only 2·6°C (36·7°F) at Bedford and 3·1°C (37·6°F) at Watnall, Notts. The noon temperature at Gatwick was at or below freezing on twenty-four of the twenty-eight days. High pressure persisted to the north or north-east, and Polar Continental air was dominant. Only the extreme west and north-west had mean temperatures above freezing. At the time of writing, there has been no complete cold month since 1986, though two short very severe spells deserve mention in the next chapter.

Fig. 88. **The persistent cold of February 1986.** The daily maximum and minimum at Kew Gardens.

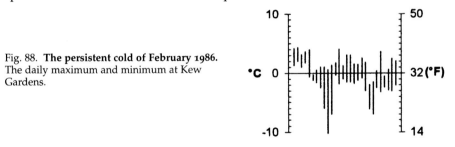

Summer Heat[5]

The effects of great heat are perhaps less sensational than those of extreme cold. Frost usually covers the landscape with ice and snow, but a hot summer's day of 30°C (86°F) differs little in appearance from an ordinary one of 20°C (68°F) except that long-continued heat may wither or brown the leaves of some trees and produce a generally dusty appearance in the landscape.

In 1911 the heat persisted throughout July, August and the first half of September (Fig. 89). The thermometer reached 27°C (80°F) on about forty days in central London. The average maximum was 27·6°C (81·7°F) in July and 27·1°C (80·8°F) in August, almost 5°C above average. The absolute highest temperature was 37·8°C (100°F) at Greenwich, but this reading is not now accepted as the thermometers were exposed in an open-fronted stand; nor is the reading of 38·1°C (100·5°F) at Tonbridge, Kent, on 22 July 1868. The summer of 1933 was persistent rather than hot, though 34°C (93°F) was recorded at Cambridge and Margate on 27 July. June 1940 was notably warm in the south. The high pressure gave calm conditions, and smooth seas helped with the evacuation of tens of thousands of British soldiers from the beaches of Dunkirk.

Fig. 89. **The daily maximum and minimum at Kew in the hot summer of 1911**

August 1947 produced Mediterranean-like conditions and average maximum temperatures were just below those of August 1911: 27·2°C (81°F) at Shinfield near Reading; 26·7°C (80·1°F) at Worcester; 25·4°C (77·7°F) at Norwich; 26·2°C (79·2°F) at Bournemouth; 23·4°C (74·1°F) at Tenby; and 23·2°C (73·8°F) at Armagh. At Fort Augustus, the average maximum of 22·8°C (73°F) was above the normal for London. Because of persistent east winds, places near the North Sea were much cooler. At Gorleston and Margate, the average afternoon maximum was only 20°C (68°F) although abundant sunshine compensated for the lower temperature.

September 1949 was nearly 5°C (9°F) above average at Kew and Ross-on-Wye. July, August and September of that year produced a very good summer. July 1955 was a very warm month with an average daily maximum of 24·7°C (76·5°F) at Falkirk and 25·2°C (77·4°F) in London. The monthly mean for the Isles of Scilly was 19·4°C (66·9°F), worthy of the Loire valley in France. August 1975 was not quite as warm as August 1947 and the heat and drought were nowhere near as widespread, but the really hot weather was probably more intense. At Kew and Birmingham the average maximum was 25° and 24°C (77° and 75°F) respectively. In Northumbria, however, this month was probably one of the best on record, as the wind blowing from the south or south-east prevented the formation of sea-breezes and stopped the landward spread of fog or low cloud from the North Sea. In general, it was an excellent holiday month for those near the sea, but the first ten days of the month were uncomfortably hot for those still at work.

June, July and August 1976 saw a repetition of weather which, for many, was again uncomfortably hot. June 1976 was the second warmest since 1815 at Oxford, and for a century at many places from Essex to Devon. The average maximum exceeded 25°C (77°F) as far north as Nottinghamshire. July was even warmer, with an average maximum of 27°C (81°F) in the London area and 26°C (79°F) in Oxford. While there were some high night minima, mostly in towns, the dry air allowed intense radiative cooling at night, particularly in the country. Ground frosts were even reported at a number of valley sites and at Kew. August was a little cooler in the south-east than the preceding two months. Again, there were average maxima of 25°C, notably in north-west England, but also some ground frosts. An air minimum of –0·6°C (30·9°F) was recorded at Great Gaddesden, Hertfordshire, on 1 August. For most of June a ridge from the Azores high dominated the weather; however, fronts affected the north-west where sunshine was well below average, giving a distinctly poor summer month in the Western Isles. Later in the month and during July, pressure was high over or near the British Isles. By August the anticyclone was located just to the west of Ireland or over Scandinavia, giving a spell of easterly weather and making parts of the west the hottest districts. Daily temperatures at Kew are shown in Fig. 90.

Fig. 90. **The outstanding summer of 1976.** The daily maximum and minimum at Kew during June, July, August and the first nine days of September.

According to the Central England Temperature (CET) series which gives the temperature at a selection of places in the Midlands as far back as 1659, as we have seen, July 1983 was the warmest of any month on record. It is not held in the popular memory like 1976, because the amount of sunshine was less than in 1976 and there was more rain. So 1976 overshadowed 1983. In 1976 there were sixty-two days with a maximum of 25°C or more, 1983 could only achieve twenty-two. Parts of the east coast and the extreme north did not share the warmth. The average maximum at Heathrow was 27·6°C (81·7°F) and 25·5°C (77·9°F) at Elmdon, 24·3°C (75·7°F) at Ringway and 22·1°C (71·8°F) at Aldergrove.

July and August 1990 produced some very hot weather. From 11 July until the end of the month temperatures exceeded 26°C (79°F) daily somewhere in Britain. Heathrow recorded 32·5°C (90·5°F) on the 21st. Sunshine was abundant and exceeded ten hours a day around the southern coasts. Parts of western Scotland had below-average temperatures as the high pressure over the south of England gave westerly winds off the Atlantic and more cloud. The beginning of August saw Britain's highest temperature of 37·1°C (98·8°F) at Cheltenham.

The summer of 1994 was a poor one in the west, with below-average sunshine in Cornwall, for example, but the south-east fared better, as did Manchester, sheltered from south-west winds by the mountains of Wales. There were spells of abundant sunshine and high temperatures, 21°C (70°F) at the tip of the Hebrides and sixteen hours of sunshine in one day at Fair Isle, but also numerous heavy downpours in various parts of the country. The daily maximum average for July was 26·2°C (79·2°F) at Heathrow; 25·8°C (78·4°F) at Cambridge but only 17·2°C (63°F) at Valentia. Eastern England was just on the edge of another very hot summer. In mainland Europe the afternoon maximum was 28·3°C (82·9°F) in Paris, some 4°C (7°F) above average. Berlin had 29°C (84·2°F), 5·3°C (9·5°F) above average, and Stockholm 27·4°C (81·3°F), 6·3°C (11·3°F) above average. So Britain was in the zone of conflict between very hot continental air and cooler air from the Atlantic.

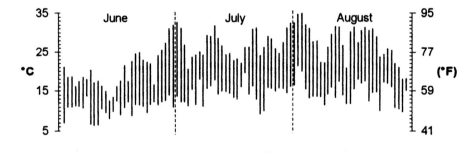

Fig. 91. **The summer of 1995 at Kew Gardens.** The daily maximum and minimum.

July and August 1995 produced a brown or yellow-brown landscape in many parts of England for there was great heat, and rainfall was well below average. There were hosepipe bans and further water restrictions to householders as reservoir levels fell. By mid-September the grass was green and lush, such is the power of recovery of grass. Apples, plums and pears seem to have done well, but vegetable crops fared badly. The highest reading in August was 34·3°C (93·7°F) at Heathrow. Cape Wrath in the extreme north-west of Scotland recorded 27°C (80·6°F). At Kew there were thirteen days with maxima above 30°C (86°F). August was among the hottest, driest and sunniest on record (see also Chapter 10). The Central England Temperature (CET) was 19·5°C (67·1°F), not far behind July 1983.

People often talk of a summer as a whole as being good or bad, depending on how lucky or unlucky they were with their holidays. In an attempt to measure a summer more objectively a number of people have developed the idea of an index to combine such basic elements in the weather as temperature, sunshine and rainfall. For the first two elements points are added, while rainfall is considered less desirable and is given a negative total. The higher the index figure the better the summer. One of the first such analyses was put forward in 1962 by R. M. Poulter.[6] Using his index, the ten best summers between 1880 and 1994 at Kew were 1976, 1911, 1899, 1983, 1994, 1933, 1975, 1949, 1921 and 1984. The worst were 1888, 1903, 1890, 1956, 1954, 1920, 1912, 1927, 1891 and 1894. Three summers, 1968, 1977 and 1980, just fail to make the worst ten.

The Poulter Index uses mean temperature, total rainfall and sunshine for the three summer months of June, July and August. A similar index, the Optimum (Summer) Index,[7] uses average maximum temperature for the same period instead, on the basis that afternoon temperatures make the most impression on people. Using this way of measuring a summer, 1976 was the best since 1880 at a number of widely scattered places in England and Wales. At Stornoway both 1955 and 1968 were better than 1976, which goes to show that there is no ideal place for high pressure to appear for all parts of the British Isles to have really good weather at the same time, as was mentioned in Chapter 3. Over Ireland and west Scotland and the Hebrides the summer of 1995 ranked some way above 1976. In the Channel Isles the summer of 1989 was much better than 1976 or 1995.

Measuring poor summers in the same way, 1888 was the worst at Kew and Southampton; 1954 at Stonyhurst; and 1912 at Edinburgh and Plymouth. There is yet another way of measuring a summer: by the amount of sunshine. Details are given in Chapter 18.

Winter Mildness

This is less spectacular than winter cold. Some notably mild months include December 1934, probably the mildest December on record, with a central England mean of 8·1°C (46·5°F). There was no temperature below 5°C (41°F) at Falmouth, while at Wick in the north of Scotland the December mean was a mere fraction below the normal for May. December 1974 was not far off a record. At Kew, the thermometer reached 10°C (50°F) or more on twenty days in the month, and at Falmouth on twenty-one. Tropical Maritime air was dominant. At the Royal Horticultural Society's Garden at Wisley, in Surrey, the rose beds had a colourful display right through the month and on into January 1975, which was also a very mild month in southern England (though Januarys 1921 and 1932 were milder). High pressure over Spain and the Bay of Biscay continued to feed Tropical Maritime air across the country, even if incursions of polar air into Scotland made it less spectacularly mild there; all the same, some of the Scottish ski resorts had no snow. In Bushey Park, near Hampton Court, daffodils were in full bloom on 16 January. Many places did not report the first air frost of the winter until the 19th. At Southampton, there were twenty-one days over 10°C and even eleven at Acklington on the Northumberland coast. The Januarys of 1983 and 1990 were only a tenth of a degree behind 1975.

February 1961 gave record warmth. Temperatures were so high that warmth not mildness was the word. A notable feature was several days of brilliant sunshine. On St. Valentine's Day, Bromley recorded 18·3°C (65°F) and Largs in Ayrshire 15·6°C (60°F). March 1938 and 1945 were both phenomenally warm in Scotland and Northern Ireland. At Edgbaston, 15°C (59°F) was reached on seventeen days in March 1948, a value more typical of May.

The 1989–90 winter was the second mildest on record. Pressure was high over central Europe so that mild Atlantic winds were dominant. On 28 December Lossiemouth recorded 15·6°C (60·1°F). Aviemore had no lying snow at 9 h in January and at many places there was no air frost. February was also very mild, with 15·8°C (60·4°F) in London early in the month. The dominance of westerly winds meant that the west had an exceptionally wet winter. There were some notable storms and a record gust of 222 k.p.h. (138 m.p.h) at Fraserburgh. London, Birmingham and Doncaster recorded 18°C (64·4°F) over the half-term week. In Kent there were several days of light winds, clear blue sky and a warm feel to the air. March was mild and dry; but April saw damaging frosts, partly due to the advanced state of many plants.

Summer Cold

June 1971 shows just how chilly a summer month can sometimes be. At Watnall, Nottinghamshire, the maximum temperature was 8·9°C (48°F) on the 9th, which is colder than many a January day. At Gorleston in Norfolk and Stonyhurst in Lancashire the maximum was 15°C (59°F) or less on seventeen days in this month; at Cambridge there were nine and at Kew seven such days. In many parts of the country the highest temperature failed to reach 21°C (70°F). June 1972 had a lower mean temperature in many places than June 1971. There were six ground frosts at Kew and ten at Exeter.

The first eight days of July 1978 were also very cool. At Manchester (Ringway), the temperature remained below 15°C (59°F) until the 9th. These cool conditions were accompanied by a strong breeze. In Surrey the audience watching an open-air performance of a Shakespeare play needed their winter clothes to keep warm, even at the matinée, for the wind was force 4 and the temperature just over 10°C (50°F).

Fortunately, such cool days are not very common. An analysis from Bidston Observatory[8] on Merseyside shows that there have been a mere six Julys since 1875 which have had daily maxima below 13°C (55°F). These were in 1978, 1954, 1948, 1930, 1902, and 1888. Altogether eight July days since 1875 (to 1982) have had maxima under 13°C.

June and July 1980 were two very poor summer months. The mean average was about 1° to 2°C below normal over many parts of the British Isles; but a better indication of high summer in 1980 is provided by the average from 18 June to 18 July. The mean maximum at Kew for this period was 16·1°C (61°F), some 4°C (7°F) below average. On seven of these days the thermometer failed to reach 15°C (59°F) and these cool days felt even chillier as the wind was often strong. In west Wales at Brawdy, Dyfed, the maximum temperature was below 15°C on nineteen days in June and thirteen in July. Many people felt that 1980 was producing the worst summer on record, but if we use an index which measures maximum temperature and sunshine[9], the three months June, July and August were not the worst, at least in the south-east of England. At Kew in the hundred years 1881–1980, there were twenty-seven summers which, according to the index, were worse than 1980. Notable among them were 1888 and 1954, followed by 1890, 1894, 1927, 1956 and 1978.

June 1991 was a very dismal month. Rain fell on twenty-eight days along the Sussex coast and the thermometer was not above 16°C (61°F) at one o'clock in the afternoon on any day during the whole month at Crawley in inland Sussex.[10] Between the 1st and the 4th, and around the 18th, much of Scotland and the north of England had maxima of 8° to 10°C (46° to 50°F), lower than many a January day, and feeling much colder because of the evaporative cooling brought about by cold winds, whereas the winter temperatures of 10°C are usually associated with high humidity and little evaporative cooling. Air frosts were recorded at Durham (−0·6°C, 30·9°F); Finningley (−0·6°C); Elmdon (−0·8°C, 30·6°F), and in the New Forest, Hampshire. In Liverpool, there was much damage to bedding plants, especially to dahlias and busy lizzies (impatiens).

The first half of June 1995 was distinctly chilly: 12° and 13°C (54° and 55°F) on the 13th and 14th, for example, with quite brisk north-east winds coming off a cold North Sea. The rest of the month was warm or hot, except for one day, the 25th, when stronger than expected winds off the North Sea made a very chilly day, at least in the eastern half of the country.[11]

In May 1996 the mean maximum CET was 13·4°C (56·1°F), 2·1°C (35·6°F) below normal, and the mean minimum CET was 4·8°C (40·6°F), 2°C (35·6°F) below normal. This gives an overall CET of 9·1°C (48·4°F) and there have been twelve colder Mays in the last 337 years, notably 1698 with 8·5°C (47·3°F) and 1740 with 8·6°C (47·5°F). However, most of these dire months occurred during the 1700s and 1800s; during the twentieth century there has been only one significantly colder May, in 1902, although the Mays of 1923 and 1941 were almost as cold.[12]

NOTES TO CHAPTER 13

1. G. Manley, Q.J. 100, pp. 389–405.
2. J.E.B. Raybould, 'A study of cold winters in the Manchester area', W. 18, pp. 275–7 (updated from M.W.R.).
3. W. (1947) 2, p. 95, and M.W.R.
4. L.S. Laskey, 'The winter of 1978–79 on the south coast of England', J.M. 4, pp. 146–7.
5. M.W.R.
6. R.M. Poulter, 'The next few summers in London', W. 17, pp. 253–5. G.S. Varney (updating of Poulter Index), C.O.L. Oct. 1976, No. 78.
7. N.E. Davis, 'An optimum summer weather index', W. 23, pp. 305–17. A.J. Cutler, 'Seasonal indices, a review', W. 28, pp. 59–63. T. Baker, 'A simple summer weather index', W. 24, pp. 277–80. R.E. Blair, 'An index of Scottish summers', W. 35, p. 39.

8. N. Hanson 'Low temperatures in July. Past and present', C.O.L. July 1978, No. 99, p. 10.
9. I.T. Lyall, 'A study of British summers using an index. (i) Trends and cycles', J.M. 1, pp. 183–7.
   I.T. Lyall, ' Recent summers and future ones', J.M. 5, p. 222.
10. Noon temperatures at Crawley. Weather Log, June 1991.
11. G. Spellman, 'Summer 1995–Britain's Best Yet?', J.M. 20, pp. 384–91.
12. *Sunday Telegraph*, 2 June 1996.

14 | Sensational Days

This chapter describes unusually hot or cold days.

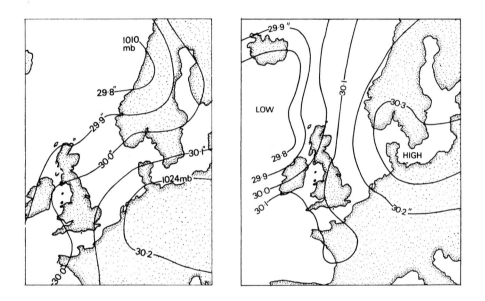

Fig. 92. (*left*) **Synoptic chart for 7 h, 1 September 1906.** Between 31 August and 3 September 1906 the thermometer reached 32·2°C (90°F) or more.

Fig. 93. (*right*) **Synoptic chart for 7 h, 9 August 1911.** An anticyclone centred over north Germany and the Baltic fed Tropical Continental air over Britain.

1 September 1906 (Fig. 92)

Between 31 August and 3 September 1906 the thermometer reached 32·2°C (90°F) or more. Pressure was high over Germany, and warm southerly winds brought tropical air. At Bawtry, Notts; Barnet, Herts; and Epsom, Surrey; 35·6°C (96°F) was recorded on the 2nd. In Scotland the thermometer reached 31°C (88°F) at Paisley.

9 August 1911 (Fig. 93)

An anticyclone centred over north Germany and the Baltic fed Tropical Continental air over Britain. The highest temperature recorded this century in an orthodox Stevenson screen occurred at Ponders End, London,[1] where the thermometer slightly exceeded blood heat, reaching 37·1°C (98·8°F). Although the screen was standard other factors have since disqualified the reading. The same value was reached in Cheltenham in 1990.

1 April 1917 (Fig. 94)

Low pressure over Scandinavia and high pressure west of Iceland brought very cold Arctic air across Britain. On 1 April the maximum temperature was actually below freezing at Tynemouth, Aberdeen and Harrogate. At Harrogate the maxima for the first five days of April were: −0·4, 1·6, 4, 4 and 7°C (31·3, 34·9, 39·2, 39·2, and 44·6°F). The minima were even more remarkable falling to −10°C (14°F) on the 2nd and to −6°C (21°F) on the 3rd. Other very low readings included: −15·5°C (4°F) at Eskdalemuir; −15°C (5°F) at Newton Rigg, Cumbria; −14°C (7°F) at Worksop, Notts, −13°C (9°F) at Balmoral, −12°C (10°F) at Burnley, Wakefield and Belvoir Castle. In the south of England the cold was not severe and Kew did not report a frost on that day.

18–21 August 1932 (Fig. 95)

High pressure extended across Scotland from an anticyclone on the Atlantic towards another over Germany, while a shallow depression lay over the Channel and south-west England. At Kew 31·6°C (88·9°F) was recorded on the 19th and 33·3°C (92°F) on the 20th. At Oxford and Reading 35°C (95°F) was recorded, at Guernsey 31°C (88°F). Minima exceeded 21°C (70°F) on two successive nights in London, but by the 22nd cooler north-east winds had spread across bringing maxima down to 19°C (66°F).

21 January 1940 (Fig. 96)

High pressure over Scandinavia and a strong ridge across Scotland brought very cold air. This, together with clear skies and a deep powder snow-cover, permitted intense cooling. The night of 20–21 January produced some very low temperatures, among them: −23·3°C (−10°F) at Rhayader and −21·6°C (−6·9°F) at Llandrindod Wells in mid Wales; −21·1°C (−6°F) at Ambleside, and Bodiam, East Sussex; −20·5°C (−4·9°F) at Dalwhinnie; −20°C (−4°F) at Canterbury, Welshpool, Hereford, Newport, Shropshire, and Houghall, Durham; −19·4°C (−2·9°F) at Braemar, Worcester and Whitstable; and −18·8°C (−1·8°F) at Castleton, Yorks, and Wye, Kent. The Norfolk coast and the Outer Hebrides escaped the severe frost as the chart shows.

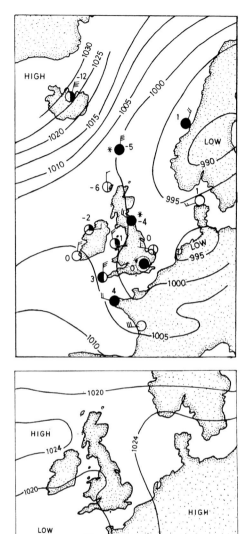

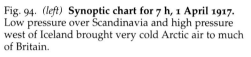

Fig. 94. *(left)* **Synoptic chart for 7 h, 1 April 1917.** Low pressure over Scandinavia and high pressure west of Iceland brought very cold Arctic air to much of Britain.

Fig. 95. *(below left)* **Synoptic chart for 7 h, 19 August 1932.** High pressure extended across Scotland from an anticyclone on the Atlantic toward another over Germany, while a shallow depression lay over the Channel and south-west England. 35°C (95°F) was recorded at Oxford and Reading in this hot spell.

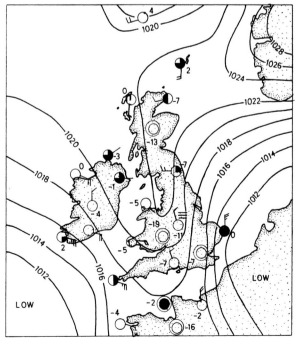

Fig. 96. *(right)* **Chart for 7 h, 21 January 1940.** High pressure over Scandinavia and a strong ridge across Scotland brought very cold air. This together with clear skies and a deep powder snow-cover permitted intense cooling: −21°C (−6°F) at Bodiam, East Sussex.

25–31 July 1943

Temperatures were very high this week reaching 32·7°C (91°F) at Kew, and 31·1°C (88°F) at North Berwick, Dunbar and Armagh. The highest readings were at Croydon and Worcester with 33·8°C (93°F).The hot spell ended abruptly with a line squall, really a sharply defined cold front. At Wigan, Lancashire, a dust cloud 30 metres (100 feet) high preceded a storm of heavy rain. At West Kirby, Merseyside, 25·9 mm (1·02 in) of rain fell in sixteen minutes. At Kew and Bidston, there were gusts of 100 k.p.h. (62 m.p.h.).

19–29 January 1945

There was continuous frost in many places away from the west coast during this period for seven out of the ten days, and there were some extremely low maxima, among them –9·4°C (15°F) at Leeming, Yorkshire, and –7·7°C (18°F) at York on the 24th, and –7·7°C at Hawarden, near Chester, on the 26th, giving a severe daytime frost. They are among the lowest maxima recorded at lowland weather stations in England this century. Throughout the period, pressure was low to the east, and small disturbances moved south or south-east across the country. The cold spell ended with heavy snow turning to rain. The coldest night and warmest day of the month were separated by only ninety-six hours.

February 1947

A very dull month, particularly over England (though west coasts were sunny); there was an unusual combination of dull skies, wind and frost. Only towards the end of February did the sky clear sufficiently to permit intense cooling. Low temperatures included –21°C (–6°F) at Woburn, Bedfordshire; –19°C (–2°F) at Luton, Appleby (Cumbria), Stratford-on-Avon, Hereford, Abingdon and Braemar; –17°C (1°F), Rye; –16°C (3°F), Totnes and Perth; –13°C (9°F), Dungeness; –11°C (12°F), Garvagh, Londonderry; –9°C (16°F), Kew; –5°C (23°F), Jersey; –3°C (27°F), Cape Wrath; and –1°C (30°F), Scilly.

4 March 1947

The Houghall frost hollow, near Durham, recorded –21°C (–6°F). Other notable lows included –21°C at Peebles and West Linton in southern Scotland. In Northern Ireland the lowest reading was –15°C (5°F).

31 May–3 June 1947

Four days over 32°C (90°F) in London and the south-east. A cold front moving south-east brought cooler weather and by the 5th the maximum in London was only 18°C (64°F).

27–28 July 1947

32°C (90°F) again recorded in London. The night of the 28th was very stuffy indeed, with a minimum of 20°C (68°F).

16–17 August 1947

34°C (93°F) at Southampton, 33°C (91°F) at Littlehampton and 32°C (90°F) at Worthing. Earlier in this very hot month, 31°C (88°F) was recorded in Scotland at Kilmarnock.

9 March 1948

23°C (73°F) at Kensington Palace, 22°C (72°F) at Greenwich, just exceeding the record established in 1848.

July 1948

This month produced a short, intensely hot spell with three successive days over 32°C (90°F), culminating in a record for the month of 34°C (93°F) on the 28th at Kew. On the 29th, 32°C was recorded at Prestwick, a rare event for Scotland. The first twelve days of this month were very cool with daytime maxima under 16°C (61°F). The contrast made the sudden heat seem very oppressive.

1949

16 April was the peak of a very warm Easter with temperature and sunshine figures which would be very creditable for August Bank Holiday in many parts of southern England, 29°C (84°F) being recorded in London. This hot spell was the first of several during a very good summer. Worcester recorded 33°C (91°F) on 12 July. Maldon, Essex, recorded 33°C on 5 September, while summer temperatures were still occurring in October. On the 3rd, 26°C (79°F) was registered at Camden Square, London. The night of 14–15 October was notably warm with a minimum of 17°C (63°F), equalling the 1921 record.

29 March 1952

Along parts of the south coast maxima were below 0°C, probably the latest date in the year for such an event. Snow fell all day at Worthing. Accumulations there were not great, but the South Downs were well covered. There were deep drifts in the Haslemere area.

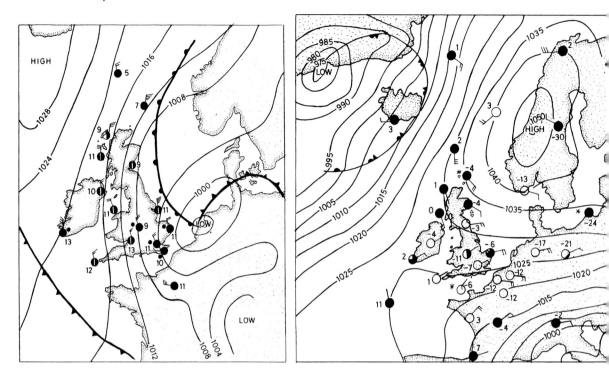

Fig. 97. *(left)* **The weather map for Coronation day, 2 June 1953.**

Fig. 98. *(right)* **Severe cold on 1 February 1956.** Very cold Polar Continental air swept across southern England on 31 January, bringing a dramatic fall of temperature from 7°C (45°F) at 8 h to –2°C (28°F) at noon, the weather changing from continuous rain to fine powder snow.

2 June 1953, Coronation Day (Fig. 97)

Those who waited all night on the pavement to get a good view of the royal procession will remember the chilly, dull, drizzly weather. Temperatures were at a level typical of a mild January.

31 January 1956 (Fig. 98)

Very cold Polar Continental air swept across southern England, bringing a dramatic fall of temperature from 7°C (45°F) at 8 h to –2°C (28°F) by noon, at Washington, just north of Worthing, the weather changing from continuous rain to fine powder snow.

Many cars were immobilised as the rain froze, jamming the doors. On the 1st, Ipswich and Felixstowe had maxima of –6°C (21°F). Kew and Ross-on-Wye were just over a degree warmer. On the 2nd, only Benbecula, extreme western Ireland and Scilly exceeded freezing point.

18–25 February 1956

Many places in the south had continuous frost which penetrated the ground deeply as, except along parts of the east coast, snow-cover was light and incomplete. Winter wheat suffered from frost heave as there was hardly any snow-cover.

28–30 June 1957

During this hot spell, 35°C (95°F) was recorded at Heathrow. A number of places in the south-east recorded 32°C (90°F) during the three days. It was the climax of a warm month. At Kew, the thermometer reached 27°C (81°F) on ten days, although with high pressure to the north or north-west the east coast did not share the warmth.

24–26 January 1958

A spectacular rise in temperature occurred as very cold northerlies gave way to Tropical Maritime air. On the morning of the 24th, –19°C (–2°F) was recorded at Shawbury, Shropshire; –19°C at Driffield, Yorkshire; –14°C (7°F) at Bristol; and –13°C (9°F) at Haverfordwest. By the 27th, Fort Augustus and Llandudno recorded 16°C (61°F).

14 March 1958

Over deep snow-cover, Logie Coldstone in Grampian (Aberdeenshire) recorded –23°C (–9°F), the lowest March temperature on record in the British Isles. Other low readings during the cold spell included –22°C (–8°F) at Grantown-on-Spey, at Braemar—also at Corwen (Clwyd) on 3 March.

1959

The last two days of February gave unusual warmth in widely scattered places. At Ross-on-Wye, and Banff in north-east Scotland, 16°C (61°F) was recorded; at Greenwich 19°C (66°F). In retrospect, it was a foretaste of a very good summer in the south. The peak of the summer heatwave occurred on 5 July, when 36°C (97°F) was reached at Gunby, Lincolnshire; 34°C (93°F) at Cromer; and 32°C (90°F) at Herne Bay, Kent. In August, the highest reading was 31°C (88°F) at Finningley near Doncaster, on the 20th; Gatwick Airport recorded 30°C (86°F) on 11 September, a noteworthy reading, but some way below the 32°C (90°F) in London in 1926. The first week of October was very warm with 27°C (81°F) exceeded from Sussex to Yorkshire. Among the highest were 28°C (82°F) at Faversham and Mickleham, Surrey.

Christmas 1961 (Fig. 99)

This was one of the coldest Christmases of the century, with bright sunshine and biting cold easterly winds. There was widespread white frost but no snow, so 1961 did not count as a white Christmas.[2]

1–2 June 1962

At Guildford the author observed from his garden at sunset a beautiful clear sky with a cold Arctic tinge. The morning of 2 June was frosty and the locals in a Midhurst, West Sussex, shop on a bright Saturday morning were loud in their complaints about damaged tubers. Air frost occurred at Rustington near Worthing. Other lows included –1·7°C (29°F) at Mickleham and –5·6°C (22°F) at Santon Downham on 1 and 3 June. It is strange to think that a short June night can achieve readings which would be worthy of comment in mid-winter, in the centre of any of our big cities.

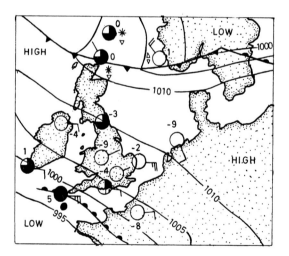

Fig. 99. **Christmas 1961.** A very frosty Christmas, with high wind chill. This cold spell lasted into the New Year. There was widespread snow on 31 December.

18 January 1963

Fig. 100 shows the pressure pattern on a very cold day in the winter of 1963. The thermometer did not rise above freezing for another eight days at Kew. Iceland was once again mild, when cold affects Britain from the east.

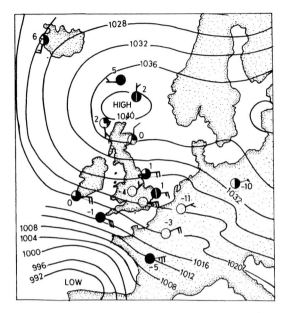

Fig. 100. **12 h, 18 January. A very cold day in the severe winter of 1963.**
The thermometer did not rise above freezing for another eight days at Kew.

28 August 1964

A southerly airstream on the flank of an anticyclone over Germany brought high temperatures. Cromer's reading of 33°C (92°F) was the highest in a good but not hot summer. As the Poulter Index shows (page 155) the 1960s were not favoured by hot summers.

29 March 1965

Wakefield and Whitby recorded 25°C (77°F). A relative humidity of 4% was recorded at Ringway and on Great Dun Fell, in the Pennines. Huddersfield recorded 10% and Aberporth and Tynemouth 11%. These low humidities are likely to be caused by descent of air from a great height; an example of subsiding air in an anticyclone, which was able to reach the surface.

19 January 1966

A clear sky, light winds and snow-cover in Kent permitted intense radiation; Elmstone recorded −18.9°C (−2°F) and East Malling −16°C (3°F). The cold was not so severe in Scotland.

15 February 1966
While many places along the south coast and in central London did not report an air frost at all, conditions in Scotland were severe with –20°C (–4°F) at Balmoral, and –17°C (1°F) at Turnhouse, near Edinburgh.

21 July 1966

Perth in Scotland recorded the maximum, 29°C (84°F), for the month on the 21st. South-eastern districts which often produce the records were covered by cold east-erlies blowing round low pressure.

29 March 1968

25°C (77°F) at Santon Downham and Cromer, equalling the record for the month. This warm spell from the 25th to 29th, unlike some other spring heat-waves, was not an indication of a good summer to follow.

16 February 1969

Among low readings were: Newton Rigg, near Penrith –20°C (–4°F); Galashiels –19°C (–2°F) and Houghall –18°C (0°F). Earlier in the month, Manchester had its coldest February night of the century with –13°C (9°F).

28 January–1 February 1972 (see Fig. 14)

Places in the south-east of England had seventy-two hours of continuous frost, though neither Hartland Point nor Plymouth had any frost during the month. In the east there were widespread low temperatures: –18.5°C (–1°F) at Moor House, 561 metres (1840 feet), on the Pennines; –18.3°C (–0.9°F) at Corbridge; –17.7°C (0.1°F) at Warsop, Nottinghamshire, and East Malling, Kent; –17°C (1.4°F) at Turnhouse, Edinburgh; –15.6°C (4°F) at Gatwick Airport; and –15°C (5°F) at Fernhurst, West Sussex. At Cromer and Binbrook, Lincolnshire, the thermometer never rose above –6°C (21°F) on the 31st. By the evening of 1 February, the Siberian high had retreated before the onslaught of a depression moving north-east near Iceland, replacing the Polar Continental air by mild conditions from the Atlantic.

3–8 August 1975

During this hot spell new records were established. The reading of 31°C (88°F) at Glasgow had not been exceeded since records began in 1868. In East Anglia and south-east England 34°C (93°F) occurred quite widely. Night minima over 21°C (70°F) were also common in 'the heat island' of urban areas, and many people found this meteorologically interesting spell almost unbearably stifling. The cause of it all was a ridge of high pressure from the Azores anticyclone, which developed a separate centre over England and then moved gradually eastwards producing a hot south-easterly airstream. The north-east of England shared in the hot weather as the gradient wind (the wind as a result of the pressure pattern) prevented the formation of a sea-breeze on many days.

8–11 May 1976

There were four successive days over 25°C (77°F), which brought the highest May readings since 1953. 29°C (84°F) was recorded at Heathrow. Like the similar early May hot spell in 1947, it was an omen of things to come.

23 June–7 July 1976 (Fig. 102)

On 21 June a weak warm front travelled north-east across the country and high pressure extended across the British Isles behind it, beginning a memorable hot spell, which proved to be the longest and hottest ever recorded in many parts of the country away from the north-west. The author's thermo-hygrograph chart for part of the hot spell is shown in Fig. 101. It was accompanied by a daily average sunshine total over fourteen hours a day in Essex and east Kent, falling to twelve hours over Dorset, the Welsh border and the Cheviots. Even in the leafy suburbs, the night minimum remained above 20°C (68°F), indicative of stifling heat in our concrete conurbations, and indoors generally. At Southampton, there was a maximum of 35·6°C (96°F). On 30 June many places recorded low humidities, such as 18% at Weybridge. On this day, visibility was very good (up to late afternoon in the Guildford area) until smoke from the numerous heath fires was trapped beneath the anticyclonic inversion, producing a thick acrid haze. The low humidity sucked the moisture out of vegetation so that fires were easily spread.

Jubilee Day, 6 June 1977 (Fig. 103)

The fine warm spell of the preceding weeks gave way to chilly, blustery weather on the 4th and 5th. The elements on Jubilee Bonfire Night will be remembered by many as being tempestuous with driving rain. Glimpses of the other beacons in the chain were few from the hilltops in the west. Those on Scafell had to contend with snow on their boots. In the south-east visibility was reasonably good; in the Northern Isles the weather was exceptionally clear. Perhaps the rain which fell on the beacons was a blessing as the ground around many of the beacons had become tinder dry and would have caught fire easily. There was no rain during the royal procession and the sun shone fitfully, though there were heavy showers in the south-west.

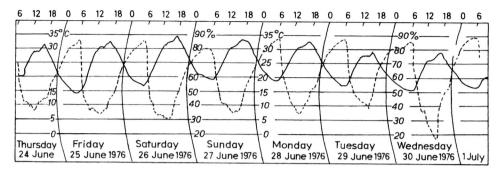

Fig. 101. **Thermo-hygrograph trace from Weybridge for the first week of the prolonged hot spell in the summer of 1976.** The small figures at the top: 0, 6, 12, 18 refer to the time; the dotted line shows the percentage relative humidity; the continuous line shows the shade temperature. Because the arms which hold the nibs which draw the lines on the chart are of different lengths (or they would clash), the temperature trace is three hours fast. Note the high reading of just under 35°C (95°F) on 26 June and the night minimum of 19·5°C (67°F) on the 27th. On the 29th and 30th, the temperatures were virtually the same, but the much less humid air made the latter day seem much cooler and fresher by contrast, 18% humidity against 38%.

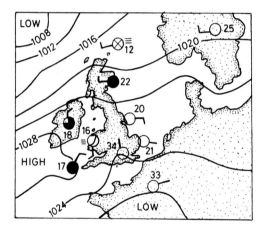

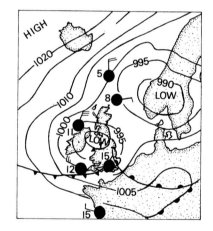

Fig. 102. (*left*) **Synoptic chart for 18 h, 26 June 1976.**
Temperatures exceeded 32°C (90°F) over many parts of England. Note the low temperatures and fog in the west.

Fig. 103. (*right*) **Jubilee Day weather, 6 June 1977.** The fine warm spell of the preceding weeks gave way to chilly, blustery weather on the 4th and 5th. Jubilee Bonfire Night will be remembered by many as being tempestuous with driving rain.

New Year's Day 1979

Places as far apart as Exeter and the Fens had maximum temperatures as low as −5°C (23°F).

3 January 1979

At Stretton, near Burton-on-Trent,[3] a minimum of −16°C (3°F) was followed by a maximum of −11·5°C (11°F). Freezing fog lasted all day.

14 February 1979

A noon temperature of −3°C (27°F) at Tynemouth was accompanied by a 77 k.p.h. (48 m.p.h.) wind, giving a very bitter day.

12–13 December 1981 (Fig. 104)

Deep powdery snow, clear sky, little or no wind and long nights combine to make Arctic Canada very cold. Such conditions occurred in wide areas of the Midlands in December 1981. At Shawbury, Shropshire, on the 12th, a new record of −23°C (−9°F) was established. With dense freezing fog, the temperature struggled up to −12°C (10°F), giving a severe daytime frost. By 18 h, the temperature was down to −22°C (−8°F), and it subsequently fell to −25°C (−13°F), beating the previous day's record. Among other low readings were Bedford −15°C (5°F); Oxford −16·1°C (3°F); Rothamsted −17°C (1·4°F); Hawarden −17·2°C (1°F); Stratford-on-Avon −19·1°C (−2°F); Newport, Salop −19·2°C (−3°F); Brize Norton, Oxon −20·9°C (−6°F); Shrewsbury −22·8°C (−9°F); and Corwen −22·8°C (−9°F). A temporary incursion of milder air affected the south as a depression moved along the Thames Valley. Fig. 104 shows the distribution of the very low temperatures.

7–14 January 1982[4]

More severe frost occurred between these dates with seven days or more of continuous frost. A number of records were broken. Among the most sensational values were: −23°C (−9°F) at Braemar on the 7th; −26·7°C (−16°F) at Grantown-on-Spey on the 8th; −27·2°C (−17°F) at Braemar, equalling the British record, on the 10th; −26·1°C (−15°F) at Newport, a new English record, on the 10th; −23·3°C (−10°F) at Newport on the 11th; −21°C (−6°F) at Elmdon and Brize Norton; and −17°C (1°F) at South Farnborough on the 14th. Elmdon had −15°C (5°F) on the 15th. Among notably low maximum temperatures were the following: −13°C (9°F) at Tummel Bridge on the 7th; −10°C (14°F) at Benson, near Oxford, on the 13th; −9°C (16°F) at Shawbury on the 11th; and −8°C (18°F) at Wyton, near Bedford, on the 14th. At the time of minimum temperature (at Elmdon) the Lichfield TV mast indicated +2°C (36°F) at 1300 feet, an inversion of 23°C (41°F).[5] Milder air from the Atlantic was sliding in over the very cold air. Subsidence of air from very high up so that the descending air was warmed by compression may be an added explanation. Shawbury's record of −25°C (−13°F) a month before did not last long.

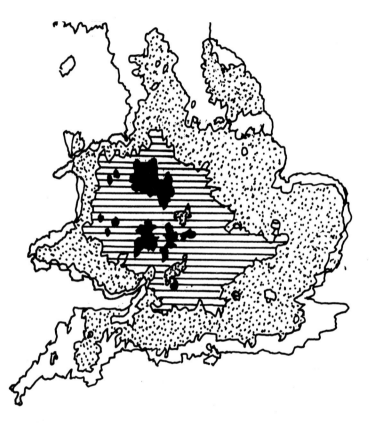

Fig. 104. **Intense cold in the West Midlands in December 1981.** This map is based on a satellite image from the National Remote Sensing Centre at Farnborough. Clear skies enabled temperatures near the ground to be measured.

A deep snow-cover, very dry air and light winds combined to produce some phenomenally low temperatures. The solid black shading shows temperatures below −15°C (5°F). In fact, a new English record was established at Shawbury, Shropshire, on 13 December of −25°C (−13°F). It will be noted that the largest areas of intense cold are in the upper Severn Valley around Shrewsbury where cold air draining downhill from the Welsh mountains can collect, and also in the lower Severn Valley and the Warwickshire Avon area. Moreton-in-the-Marsh had low readings as well. The horizontal lines indicate −10°C to −14°C (14°F to 7°F), covering much of the south Midlands, Berkshire, Wiltshire and Cambridge. The unshaded part had only a slight frost, −2°C (28°F) and above. Central London shows up as having only a slight frost, too. Coastal places in south-west Scotland actually had no frost. Over the Pennines readings were between −4°C and −10°C (25°F and 14°F). Over much of Kent readings were above −4°C (25°F). Some places in the south-west had no snow-cover, which kept temperatures from falling much below freezing.

12 January 1987

A clear sky, a wind from the east, plus a few centimetres of powdery snow, made this one of the coldest days of the century with a maximum as low as –8°C (18°F) at Reigate, Surrey; Okehampton, Devon; and Middleton, Derbyshire. Only the east coast, west Scotland, and Ireland had maxima above –5°C (23°F). Even in Jersey the maximum was –6°C (21°F). The only place in the British Isles to exceed freezing point was the northern tip of the Isle of Lewis. Lower but much more localised maxima have occurred in freezing fog in 1945, 1947 and 1981 and 1982. The daylight hours of the 12th were clear and sunny, though snow was forecast. The continuous sunshine had no real effect, even with light winds. The next day was almost as cold with snow falling with temperatures below –5°C (23°F). Who says it is too cold to snow? Noon readings at Gatwick from the 7th to 20th were in °C: 0, –2, 1, –1, –5, –7, –7, –3, –2, –1, –3, –3, –3, –1. The severe frost caused burst water pipes in Dartmouth and Kingswear, Devon, due to the penetrating power of the wind, when in most years frost does not achieve much more than thin ice on a puddle. Iceland was mild as cold air undergoes substantial warming as it rotates clockwise around the high pressure, passing over many kilometres of the relatively warm Atlantic Ocean.

The cold wave brought temperatures well below freezing to the Isles of Scilly. The frost and cold winds succeeded in wiping out 80% of the plants in the gardens at Tresco:[6] 25-metre (80 foot) high Norfolk Island pines and mature New Zealand flame trees, together with succulents and acacias, which formed the backbone of the mature planting in the garden, were killed, and other trees died later. However, many plants have shown unexpected powers of regeneration. The gardens were founded in 1834, which suggests that the 1987 frost was the worst in 153 years. The screen temperature of –7·2°C (19°F) beat the previous record of –5°C (23°F). Frost is very rare in the Scillies but when it comes it is usually associated with a wind, which is much more damaging, although ground frosts will occur in still weather. The islands are too small to allow the normal night-time cooling of a larger land mass. Without much wind, very cold air from eastern Europe would take a long time to reach the Scillies and would be warmed up by contact with the relatively warm sea, at 10°C (50°F) in mid-winter. The chart for 1 February 1956 (Fig. 98) is similar.

3 August 1990 (Fig. 105)

A temperature slightly in excess of normal blood heat 37·1°C (98·8°F) was recorded at Cheltenham on 3 August. This is now Britain's high-temperature record and equals the now suspect value recorded at Ponders End in 1911. On the same day Heathrow recorded 36·5°C (97·7°F) and Cambridge 36·7°C (98·1°F). Winds were light south-easterly, giving a short sea crossing, circulating around an anticyclone over northern Germany and southern Sweden, a pressure pattern not unlike that for previous heat-waves. On the 2nd, 33·8°C (92·8°F) was recorded in Manchester and 34·7°C (94·5°F) in Nottingham. Problems were experienced on roads and railways with melting tar and buckled rails. Coastguards were kept busy as more people were tempted into the sea to keep cool. Although the sea was one degree warmer than average the contrast between water and air was too great for some. In the memory of the general public the hot summer of 1990 made much less impression than that of 1976, partly because the heat was not so prolonged and sunshine was less abundant. There was also no Minister for Drought appointed.

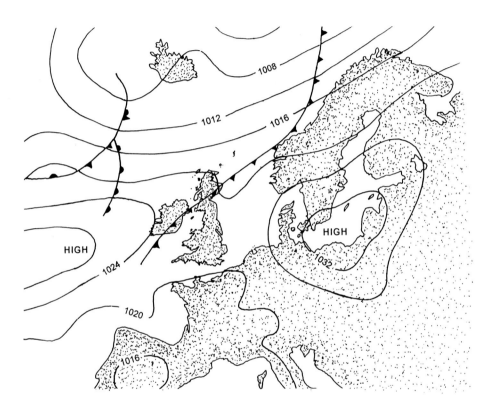

Fig. 105. **3 August 1990**. Britain's hottest day on record. 37·1°C (98·8°F) at Cheltenham. High pressure over Germany fed very hot air across much of Britain. In Scotland Leuchars reached 30·8°C (87·4°F). Aldergrove's highest of 26·7°C (80·1°F) occurred on the 2nd. The cold front moving south-east which brought fresher weather, but only a little rain in some places, spoilt Ireland's chances of achieving record heat. The front cleared the south-east of England by the 5th and the noon reading at Gatwick was some ten degrees lower than the previous day, though with light winds and sunshine. This was very pleasant.

30 July –5 August 1995

Seven successive days with maxima over 30°C (86°F) at Kew. On the 2nd and 3rd, the minima were 21°C and 20°C (70°F and 68°F) respectively. The hot weather began on 19 June and continued until 29 August. 25 June was one remarkably cold day in an otherwise hot spell with a maximum of only 16·5°C (62°F), when north-east winds and cloud from the North Sea held sway. On forty-six days the maximum exceeded 25°C (77°F); but this was not as many as in 1976 (Table 23). Pressure was high over Scandinavia and low over the Bay of Biscay, which gave a hot easterly airflow for the first few days of the month. The high then retreated westwards and the weather became less hot.

27–30 December 1995

During the second half of December Scotland was very snowy. Over deep snow-cover and with light winds and clear skies, there were four successive nights below −20°C (−4°F). In the Highlands, Aviemore recorded −24·6°C (−12·3°F); Abbotsinch (Glasgow) −20°C; and Altnaharra −27°C (−16·6°F) (a record for December; but not quite for the British Isles, by 0·2°C). On the 29th, Braemar's maximum was −15°C (5°F). The south of England was less cold: Heathrow's minimum for the month was a mere −5°C (23°F)!

NOTES TO CHAPTER 14

1. G.T. Meaden, 'Britain's extreme temperature by month of the year', J.M. 1, p. 18.
2. C. Melvyn-Howe, 'Wind chill, absolute humidity and the cold spell of Christmas 1961', W. 17, pp. 349–58.
3. D.J. Stanier, J.M. 4, p. 24.
4. National Remote Sensing Centre, Farnborough.
5. B.G. Stanley Clarke, C.O.L. January 1982, p. 14.
6. Ursula Buchan, 'Taking the lid off the Temperate House', *Daily Telegraph,* 23 July 1995.

15 | Snow

People have opinions about snow. They like it or loathe it, play with it, photograph it, or try to stay indoors and ignore it. There are others who see a centimetre of snow on, say, Wednesday morning and hastily cancel all (social) arrangements for the rest of the week. Many of our snowfalls do not last four days. There are others who won't make social engagements in the winter in lowland parts for fear of the dreaded white stuff. There are those who think snowfalls are not what they used to be. Just how much snow do we have? Is it true we have less snow now than we used to? With several snow-free winters in succession in the south recently, it is tempting to agree with those who think there was more snow in the past. But 'the past' needs definition. The 1940s, 1950s and 1960s were all snowier decades than the early 1970s; but the 1920s and 1930s were not very snowy.[1] Oxford had one snow-free winter in the 1920s, 1930s, 1950s and 1980s, and two in the 1990s, as Fig. 106 shows, while since 1917–18 Birmingham has had two snow-free winters, in 1924–25 and 1933–34. There was no snow on the ground in London and Cardiff for three successive winters, starting 1987–88.

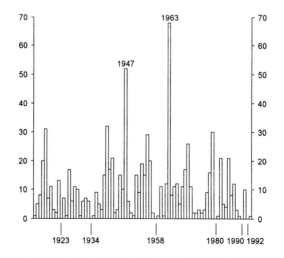

Fig. 106. **Number of mornings at 9 h with more than half the ground snow-covered at Oxford, for each winter from 1912–13 to 1992–93.** Snow-free winters indicated below graph.

The map (Fig. 107) shows the average number of days a year when snow is seen to fall.[2] The largest totals are to be found over the Highlands of Scotland, the Orkney and Shetland Isles. On low ground, Lerwick has one of the largest totals with over sixty days a year. Snowfall decreases rapidly from over sixty days near Cape Wrath to under twenty at Jura and Islay in the Inner Hebrides. Eastern Scotland is more likely to see snow, but populated areas of the Central Lowlands are sheltered from snow-bearing winds and are as likely to have snow as the Cotswolds. Over England the highest parts of the Pennines experience over sixty days with falling snow. Most of upland Wales has more than thirty, as do the hills of the Midlands and the Cotswolds. From Dyfed (Pembrokeshire) across to Essex, snow is likely on twenty days (except over the moors and Downs, where it is more frequent). Along the west coast from Anglesey round to the Isle of Wight snow is rare. Snow is least likely to fall in west Cornwall. These facts show that snow is uncommon in some parts of the country. When the definition of a day with snowfall is considered the rarity is even more apparent. Even a few flakes constitute a day with snowfall!

People associate winter with snow, but a winter can pass without a single snowflake being seen. At Penzance no snow was seen at all in two successive winters, 1942–43 and 1943–44. In Jersey the winters of 1948–49 and 1949–50 had no snow. In earlier decades, no snow fell at Bude in 1913–14 and 1922–23. Before this, no records were kept, so it is not possible to report accurately on those winters.

Of more practical interest, because of their possible inconvenience, is the average number of mornings with snow-cover.[2] Records of snow-cover are noted only once a day at nine o'clock in the morning (GMT). The observer's task is to note down any occasion when half or more of the ground representative of the station is snow-covered, a task not without its difficulties when the snow drifts. The averages vary from one day in five years at Newquay, to eight a year at Kew, nine at Wisley, sixteen at Goudhurst, eighteen at York and sixty at Braemar—but only twenty-five at Lerwick where falling snow is common. At Tiree, snow will be seen on the ground on only three mornings. Figs. 108 and 109 show maps of the average and median values of snow-cover.[3] Annual data, which are the raw material of the thirty-year averages, are shown in Table 29. Detail about snow-cover in each winter from 1939–40 to 1994–95 is shown for seventeen places in Britain.[4] Over forty years, Ilfracombe in north Devon had snow on the ground on a hundred mornings, so that the average number of mornings with snow-cover was two and a half. Yet a morning with snow-cover is decidedly rare, for nineteen winters had no snow-cover at 9 h. Six winters, those of 1939–40, 1946–47, 1962–63, 1977–78, 1978–79 and 1981–82 accounted for two-thirds of the mornings with snow-cover. Twenty-seven winters out of fifty had no snow-cover at all. It should be noted that mornings with snow-cover are not necessarily the same thing as days with snow-cover, as many snowfalls are of trifling amounts and soon melt. Sometimes snow may cover the ground for a few hours in the evening, melting away before morning. These snowfalls will not be recorded in the statistics.

At Goudhurst in the Weald of Kent, snow-cover averaged fourteen mornings a year over fifty years. The outstanding contribution of three winters is shown in Table 29. At Ross-on-Wye, eight winters have been snow-free since 1914–15, while five

Fig. 107. (*above*) **Average annual number of days with snowfall (1941–70).**

Fig. 108. (*above right*) **Average annual number of mornings (at 9 h) with snow-cover (1941–70).** Over Wales and the Midlands, snow-cover increases by about five days per 100 metres (328 feet) of altitude up to 400 metres (1312 feet). In the north of Scotland, the increase is seventeen days per 100 metres up to 400 metres. At 800 metres (2625 feet), snow-cover is probably around four times that at sea-level.

Fig. 109. (*right*) **Mornings with snow lying in a 'median' winter.** Over any number of winters, there are likely to be as many with snow- cover exceeding the median value as there are below it. In East Anglia the median value is about ten. Over twenty winters, for example, ten could be expected to have more than ten mornings with snow-cover and ten with less than this. In Cornwall and west Wales, the median value is about one, so half or more winters could be expected to be without snow-cover. The data for Ilfracombe in Table 29 support this.

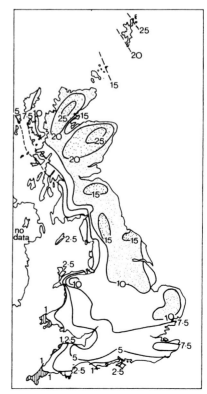

have had more than twenty-one mornings of snow-cover. On the colder side of the country, more exposed to snow-bearing winds, Waddington in Lincolnshire has had twelve winters since 1939–40 with over twenty-one mornings of snow-cover, but only two without any. At York, until 1991–92, no winter since the start of the record had been free of snow-cover; and it seems rare for a winter northwards from Lincolnshire to pass without some snow covering more than half the ground, as Polar Maritime air will usually be just cold enough for snow to settle for a short time.

Birmingham seems a snowy city by British standards. Much of it lies quite high, over 130 metres (427 feet), and astride the path of cold easterlies as well as being exposed to snow showers which penetrate through the Cheshire gap. The extra height is just sufficient to bring surface temperatures down low enough for snow to settle on a few mornings when it would not do so in other places nearby at a lower level. In 1962–63, there was snow-cover at Edgbaston on seventy-five mornings compared to only forty-five at Kew and thirty-six at Manchester. In the Scottish Highlands, Braemar's least snowy winter had twenty-three mornings with snow-cover. The snowiest winter in this part of Scotland was 1940–41, when snow-cover lasted the equivalent of fifteen weeks. In the Great Glen, 1978–79 was the snowiest, followed by 1969–70 and 1954–55. Tiree had only one morning of snow-cover in the 1962–63 winter when Birmingham had seventy-five. That island's snowiest winters were 1954–55 and 1978–79 when snow lay on the ground for seventeen mornings.

So much for snow-cover! Over a period of thirty years from 1946 to 1975, only five winters recorded a morning when there was at least 10 cm (4 in) of snow at Kew Observatory. Even in the snowiest winters there was only a fortnight when snow lay deeper than 10 cm. During the same period there was no really deep fall of 21 cm (8 in) or more. Down in Kent, there was more than 30 cm (12 in) of snow covering the ground in the winters of 1947 and 1962–63, with a depth exceeding 21 cm (8 in) on thirty mornings. At Birmingham, Edgbaston, a depth of at least 30 cm (12 in) was recorded in 1946–47; and ten winters had snow depths exceeding 10 cm (4 in). In the eastern Highlands, 40 cm (16 in) or more of snow occurred in four winters and the ground was covered with at least 21 cm (8 in) of snow for about forty days in 1946–47 and for sixty in 1962–63. In two winters (those of 1960–61 and 1970–71) the greatest recorded snow depth was under 5 cm (2 in); the winter with the longest snow-cover (1950–51) did not have more than 30 cm (12 in) of snow on any morning.

What are the conditions that favour snow? Just as we have relief, or orographic, convection and frontal rain, so it is with snow. Obviously a supply of very cold polar air is needed as well. The answer to the question 'How cold must it be before it can snow?' needs amplification. If these conditions are fulfilled, Polar Maritime air will bring frequent snow showers to the Hebrides and Highlands; sometimes a few will penetrate to the south coast of England. As was shown in Chapter 1, the warmth of the sea will warm up a shallow layer of air in contact with it so that the temperature and humidity are usually too high for snow to lie for long, as the snow-lying data for Tiree show (Table 29). Northerly winds in Arctic airstreams, with a shorter sea crossing, will often be more snowy. Everyone knows the saying: 'The north wind doth

Table 29 The Number of Mornings with Snow-Cover from 1939–40 to 1992–93

Winter of	Lerwick	Ft. Augustus	Braemar	Edinburgh	Tiree	Eskdalemuir	York	Waddington
1939–40	25	15	75	18	2	39	36	40
40–41	23	16	107	34	7	50	30	32
41–42	24	5	80	22	0	67	45	42
42–43	13	8	30	1	0	8	9	9
43–44	16	6	49	4	4	5	7	10
44–45	25	25	80	15	16	32	26	16
45–46	9	6	44	4	4	9	12	11
46–47	59	5	99	52	7	59	57	59
47–48	15	14	60	9	7	24	6	14
48–49	22	9	39	4	1	14	3	3
1949–50	16	7	18	2	1	17	8	5
50–51	35	31	102	15	9	61	17	17
51–52	39	19	55	27	2	41	8	8
52–53	16	16	58	5	3	32	11	13
53–54	15	12	36	6	0	32	18	19
54–55	46	32	86	33	17	57	24	46
55–56	43	18	65	23	4	44	40	34
56–57	10	13	31	1	1	10	2	2
57–58	44	27	72	20	11	22	28	15
58–59	28	23	60	11	2	26	12	12
1959–60	20	22	37	19	2	26	9	17
60–61	13	2	23	3	0	10	2	0
61–62	31	30	77	16	12	39	13	18
62–63	53	29	86	56	1	82	70	60
63–64	10	1	27	1	1	8	10	7
64–65	28	13	64	22	3	34	13	18
65–66	60	18	82	22	2	37	20	19
66–67	27	13	47	5	0	20	3	2
67–68	52	27	68	11	5	51	12	13
68–69	37	25	87	24	6	35	21	21
1969–70	46	36	83	13	3	35	16	27
70–71	10	5	36	1	0	19	12	10
71–72	13	8	23	10	1	16	10	9
72–73	22	11	44	6	4	28	3	0
73–74	19	21	48	7	0	16	6	6
74–75	13	12	30	1	1	15	1	3
75–76	19	15	44	2	0	12	4	2
76–77	28	24	76	11	0	44	9	8
77–78	20	26	86	25	4	34	17	17
78–79	48	53	96	30	17	62	38	43
1979–80	26	5	48	6	5	28	9	9
80–81	24	19	52	4	8	30	4	13
81–82	33	20	76	23	11	48	28	23
82–83	22	18	94	0	5	34	10	13
83–84	29	30	88	17	9	37	11	14
84–85	24	14	88	8	6	37	16	18
85–86	31	25	94	30	8	60	17	27
86–87	29	16	74	9	3	39	12	18
87–88	22	3	43	0	1	19	3	5
88–89	12	7	40	1	0	14	2	3
1989–90	15	9	38	7	1	21	2	6
90–91	8	9	56	15	7	41	16	11
91–92	12	4	30	3	1	13	0	2
92–93	15	13	51	2	2	26		4

Table 29 *continued* **The Number of Mornings with Snow-Cover from 1939–40 to 1992–93**

Buxton	Birmingham	Kew	Goudhurst	Hastings	Ringway	Rhoose	Ilfracombe	Armagh	Winter of
45	37	23	34	25	31	12	10	1	1939–40
45	35	7	22	21	17	9	1	14	40–41
57	16	11	26	10	25	1	0	9	41–42
13	3	0	2	0	2	1	0	1	42–43
27	8	0	9	5	3	0	0	0	43–44
35	19	14	15	15	13	7	7	26	44–45
17	2	3	17	7	4	1	0	6	45–46
71	59	47	55	16	49	39	19	28	46–47
32	13	8	10	7	10	1	4	13	47–48
12	4	1	1	1	6	0	0	3	48–49
7	2	1	2	0	1	0	0	3	1949–50
48	18	9	13	8	18	5	2	14	50–51
38	11	5	10	4	7	2	3	9	51–52
25	18	8	13	6	7	11	0	5	52–53
29	12	5	20	9	12	7	2	1	53–54
51	36	13	26	16	28	18	3	21	54–55
40	30	7	21	18	18	8	5	10	55–56
12	6	1	3	0	2	0	0	4	56–57
23	14	8	13	8	12	5	4	15	57–58
26	12	2	2	0	11	5	0	13	58–59
30	17	4	10	10	10	6	0	9	1959–60
10	1	0	0	0	0	0	0	1	60–61
29	15	8	13	5	13	3	1	16	61–62
74	75	45	65	54	36	51	8	39	62–63
21	12	4	18	7	1	2	0	6	63–64
34	25	5	21	10	4	5	0	13	64–65
38	22	6	13	11	5	13	1	15	65–66
14	4	0	5	1	2	0	0	3	66–67
44	23	11	18	5	13	6	0	15	67–68
50	25	6	17	9	14	4	1	25	68–69
50	34	12	24	6	16	7	1	24	1969–70
23	10	5	17	16	4	3	2	3	70–71
11	6	3	6	3	1	0	2	6	71–72
17	7	0	6	1	6	3	3	9	72–73
13	3	1	2	1	2	1	0	4	73–74
6	4	1	7	2	0	1	0	4	74–75
11	8	2	3	1	0	2	0	4	75–76
43	25	7	7	4	7	3	0	6	76–77
28	12	1	12	6	5	10	11	13	77–78
82	62	25	24	19	25	24	9	27	78–79
31	18	1	1	0	3	0	0	6	1979–80
33	19	0	6	1	3	1	0	6	80–81
35	9	23	24	9	26	17	8	20	81–82
21	7	7	9	10	4	1	0	9	82–83
25	13	0	2	0	9	0	0	17	83–84
47	33	22	32	27	19	18	6	19	84–85
60	31	13	32	20	13	3	0	13	85–86
25	24	13	12	16	9	5	0	10	86–87
9	3	0	2	0	1	0	0	6	87–88
11	4	0	0	0	1	0	0	0	88–89
14	2	0	0	0	2	0	0	4	1989–90
33	22	11	12	10	8	9	2	6	90–91
5	3	0	0	0	0	0	0		91–92
7	2	0	0	0	0	0		0	92–93

blow and we shall have snow.' Fig. 110 shows the distribution of snow-cover brought by an Arctic airstream on 10 February 1969. From mid-Sussex to Devon across to just south of Birmingham and to Dorset, there was no snow: a not uncommon event when the snow comes from the north. For Londoners the saying 'The north wind doth blow . . .' is not very reliable, since the cold air has been robbed of what little moisture it has held by the time it reaches the south, after its long passage over land. Kent, though, juts out into the wind. So do North Wales and Cornwall which on that occasion were snow-covered. This piece of weather lore would serve better in Canterbury or on the Downs above Dover than in Hampshire or Dorset.

In the south-east, about 27% of snowfall comes from northerly situations mainly affecting Kent, East Sussex and East Anglia.[5] Easterly winds account for a further 40%. Cold fronts normally contribute little snow, about 6%, unless they become stationary. For by the time the air is cold enough for the snow to reach the ground, the warm air which supplied the moisture for precipitation has usually been pushed away. A further 27% of the snowfall occurs in advance of warm fronts or occlusions. These snowfalls can sometimes be substantial, notably when warm air approaches from the south. Then three out of four fronts move into the Channel and thereafter retreat, leaving the country in cold air, while the snow falls from the uplifted tropical air-mass. This happened in 1947, 1963 and 1978. Fronts approaching from the south-west are more likely to sweep across the country, bringing milder air with snow soon turning to rain.

As a rule, for falling snow to lie on the ground the temperature must be at or near 0°C, or else the air through which the snowflakes fall must be dry, indicative of a high lapse rate. But quite often snow will accumulate when the temperature is a degree or more above freezing, if it is falling heavily. For then a snowflake as soon as it touches the ground will be covered by another before it has had time to melt. As soon as the snow falls less thickly a rapid thaw will follow.

The question 'How cold does it have to be for snow to fall?' was asked earlier. In general, snowflakes will be seen falling with rain when the thermometer stands at 2°C (36°F), giving a nasty cold sleet. Only snow would be likely at 1°C and below. Yet sometimes rain or drizzle will fall when snow might normally have been expected, even though the temperature is below freezing point. That is a comparatively rare phenomenon worthy of a special chapter (Chapter 17). The opposite occasionally occurs in the spring, as late as April or May, when large feathery snowflakes swirl about with the temperature at 4° or 5°C (39° or 41°F). Often such days seem perishingly cold, and even worse than the bleakest days of winter: it is as if the thermometer is lying, for, according to the mercury, it is too warm for snow; at the same time, from the feel of the air on hands and face, it is certainly cold enough for snow.[6]

Whether rain, sleet or snow falls will depend on the lapse rate and the humidity. The more rapid the fall of temperature with the height, the shorter the distance a snowflake will have to travel through air which is above freezing, and so the less time it will have to melt. The average lapse-rate in moist air is about 0.6°C per 100 metres (0.33°F per 100 feet) of ascent, so if the surface temperature is 2°C the freezing level will lie at 330 metres (1083 feet). If the air is dry, the lapse rate will probably be 1°C for each 100 metres (0·55°F for each 100 feet), so the freezing level will be at 200 metres (656 feet). The snowflake with only 200 metres of air above 0°C to fall through obviously has a much better chance of reaching the ground than the one which has to travel 330 metres. It is likely that the snowflake descending through dry air will have its chances of survival further increased as evaporation cools. The reader may well exclaim at this point: 'If the air is dry, how can it snow?' But that is easily answered. The air aloft can be saturated when the air near the ground is still quite dry.

'There will be rain or snow tonight', says the weather-man. It seems odd that so many forecasts have to be vague. After all, rain is common but it does not generally inconvenience us in our daily lives. The local authorities need to know whether snow will fall so that roads can be salted. Snow is very fickle at low levels in Britain, sometimes coming unexpectedly or not coming when forecast. The following example will help to explain the forecaster's predicament: very small changes in dry bulb and wet bulb temperature will make all the difference between deep snow and none at all.

Steady snow is falling with the screen temperature standing at 1°C. The surface air is fairly dry and the wet bulb temperature is −0·5°C. The ground is frozen and there are no drops of water on the end of the icicles. Had the wet bulb remained above 0°C all the time, then no snow would have accumulated. Had it remained below 0°C, all the falling snow would have accumulated, at least in theory. Once that snow has settled how long will it remain? Again this depends on temperature and humidity. Sunshine, wind and rain are other important influences.[7] Snow will remain crisp when the air is dry and the thermometer goes as high as 4°C (39°F) in the afternoon, but a rapid thaw will occur in saturated air at the same temperature.

Some people say that falling snow will not lie on wet ground. That is not strictly true since it is the air temperature which will determine whether snow will lie or not. At 0°C, a moderate snowfall will accumulate as slush on wet roofs, paths and untreated roads. Grass will often be the first surface to be covered.

'It is too cold to snow' is another saying which is widely current. As far as Britain is concerned, it is quite often true. Yet in Canada snow falls frequently down to −15°C (5°F) or lower. In January 1947, I saw steady snow falling in Brighton with the temperature at −7°C (19°F), but even so most of our snow does fall between −2° and +2°C. Often those who say it is too cold to snow are perhaps not referring to what the thermometer registers but to the feel of the air. If it feels very cold on the skin, without the thermometer registering a very low temperature, it means that the air is very dry, hence a substantial snowfall would be unlikely. In most winters, snow accompanied by biting winds does not often occur. It falls in the less cold weather which follows, or rather in the weather which feels less cold. In severe spells, Polar Continental air will sometimes produce snowfall with the temperature as low as −4°C (25°F), the snow actually falling from the uplifted tropical air, forming the thick clouds of a frontal system, or from orographic cloud as over the Yorkshire Moors and Lincoln Wolds. In Scotland, unstable Arctic air in the rear of a depression over the North Sea can also give heavy falls of powdery snow, which easily pile up into great drifts. 'It is too cold to snow' is rather wide of the mark, then, in our severe winters, though in most winters it seems to fit the facts.

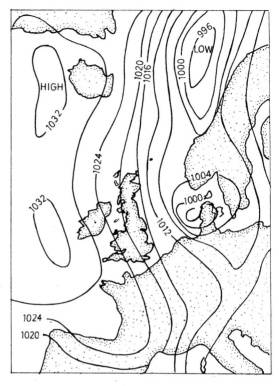

Fig. 110. (*above*) **10 February 1969.** Dots show snow-covered land. An Arctic airstream brought snow to much of the British Isles. Only central southern England, sheltered by mountains from the north wind, escaped. The synoptic situation is shown in Fig. 111.

Fig. 111. (*right*) **Noon, 8 February 1969**

So far we have described ideal conditions for snow, and where snow is common and rare. Lastly, we must answer two questions: 'When does it snow?' and 'Which is the snowiest month?' Virtually everywhere January is the snowiest month, though Buxton, Braemar and Lerwick, in the Shetlands, have more in February.[8] March is far more likely to see snow than November. Falling snow is also more likely to be seen in April than November. October snow is a rarity occurring about one year in four at Braemar. December is much less snowy than January: 1·3 to 4·3, for example, at Kew. Details are given in Table 30, also a comparison between the averages for 1941–70 and 1961–90.

Out-of-season snow covered Cross Fell in the Pennines until 10 July 1951: the previous winter was one of the worst at higher levels. The Derby was run in a snowstorm in 1888. The higher fells in the Lake District were snow covered on 16 June 1823. In 1891, snow fell as far south as London and Bath on 10 and 17 May. On 10 May 1943, there were 1·5 metre (5 foot) drifts around Moor House on Cross Fell. Snow covered the higher parts of the Pennines on Coronation Day, 2 June 1953. On 7 June 1977, the morning after the Jubilee celebrations, snow was covering the top of Skiddaw and other Lake District mountains. Perhaps the most unusual out-of-season snow occurred on 2 June 1975, even covering the ground in Edinburgh, Keswick and Birmingham. This truly remarkable event came before an abrupt change to very hot weather within a week, at the start of a hot summer.

Table 30 **Average Number of Mornings with Snow-Cover (1961–90)**

Height above sea-level (m)		Jan	Feb	Mar	Apr	May	Oct	Nov	Dec	Year average (1961–90)	(1941–70)
82	Lerwick	6·4	6·4	3·6	1·5	0·3	0·2	2·8	5·8	27·0	31·2
21	Fort Augustus	5·3	3·7	1·6	0·3	0·1	0·3	1·9	3·5	16·7	16·6
339	Braemar	15·2	15·5	9·9	3·2	0·1	0·3	3·7	11·2	59·1	60·4
23	Edinburgh	4	4	1·1				0·7	2·3	12·1	14·9
9	Tiree	1·5	0·8	0·4	0·3			0·3	1·2	4·5	2·2
242	Eskdalemuir	10·3	9·1	4	1·1			1·9	5·5	31·9	32·8
20	York	4·9	3·7	1	0·1			0·9	2·1	12·7	17·5
68	Waddington	5·1	4·2	1·7	0·1			0·7	2·5	14·3	18·1
307	Buxton	9·2	9·3	4·7	1			1·45	5	30·7	34·5
5	Kew	4·3	1·6	0·4	<0·1			0·1	1·3	7·8	9·2
85	Goudhurst	5·3	4	1·5	0·2			0·3	2·5	13·8	19·9
45	Hastings	3·8	2·3	0·7	0·1			0·2	1·2	8·3	11·3
8	Ilfracombe	0·8	0·4	0·1	<0·1				0·2	1·6	1·4
67	Cardiff	2·9	1·8	0·3	0·1			<0·1	0·9	6	8·5
76	Ringway	3·6	2	0·7	0·2			0·1	1·8	8·7	8·9
62	Armagh	4·7	2·6	1·2	0·2			0·9	2	11·6	11·7
83	Jersey	1	0·9	0·2				0·2	0·4	2·7	7

At Fort Augustus and Braemar the snow-cover has not varied significantly over the two thirty-year periods; at York the number of snowy mornings has declined by five a year and in Kent by six. In the Channel Isles there is only one-third the number of mornings with snow-cover. (An extension of the thirty-year period to include the winters of the 1990s would show a dramatic decline in snow-cover.)

In this century the earliest snowfall on the ground in England, at low levels, appears to be 31 October 1934 when 5 cm (2 in) fell as far south as Belvoir Castle, Leicestershire. Another early fall gave a light covering to the Cotswolds on 1 November 1942. Fig. 112 shows a snow patch, the remains of previous years of snow-fall lingering on the northern slopes of the Cairngorms, in a deep corrie near Ben MacDhui.[9] Observations on the summit of Ben Nevis between 1884 and 1903 gave an annual average of 169 for the number of days when snow or sleet was observed, and of over 200 for mornings with snow-cover.

Some indication of the snowiness of inhabited upland Britain over about 300 metres (1000 feet) is given by the records for Braemar. There, more than half the winters have sixty mornings (two months) of snow-cover, though it must be emphasised that the snow-cover would probably not be continuous. In spite of this, Highland Britain is snowy enough for much capital to have been invested in the ski-resorts at Aviemore, Glenshee and Glencoe. There is a ski-lift on Helvellyn, but generally over the Lake District and the Pennines the snow is not sufficiently reliable to justify much expenditure. All the same, snow-cover is likely on eighty mornings at Moor House (561 metres (1841 feet)).

Fig. 112. **The semi-permanent snow in the Cairngorms.** For many years this snow patch near Ben MacDhui was considered to be a permanent feature of the landscape and even an incipient glacier, but much to the disappointment of some people it vanished altogether in September 1933 and again in 1959; for our climate is too warm to support a permanent snowfield, or else the mountains are not quite high enough. In 1974, sixteen snow patches survived in the Cairngorms; in 1975 there were four; in 1976 just one, in 1977 thirteen. It may surprise you to know that any snow lingered on the mountain tops in 1976, which saw record-breaking heat over England. Over Scotland the summer was not as hot and sunny as 1959 or 1955. Also September 1976 produced some early snowfalls. For the snow patch to disappear a dry and warm autumn seems necessary.

NOTES TO CHAPTER 15

1. R.A. Canovan, 'Wintry prospects for British Rail', W. 26, pp. 472–91.
2. M.O., Bracknell, Climatological Memo. No. 74, 1975.
3. M.C. Jackson, 'Median snow cover in Great Britain', W. 33, pp. 298–309.
4. M.W.R. (snow cover September to June).
5. P.C. Clarke, 'Snowfalls over south-east England, 1954–69', W. 24, pp. 438–47.
6. K.O. Mortimore, 'Snow showers at high air temperatures', J.M. 2, p. 243.
7. J.M. Heighes, 'Variation of freezing point with relative humidity', J.M. 2, p. 304.
8. P.C. Clarke, op. cit.
9. P.C. Spink, 'Scottish snowbeds in summer, 1979', W. 34, pp. 158–60; also, annual survey, W. 10 to 33. D.L. Champion, 'Summer snows around Ben Nevis', W. 7, pp. 180–4. I.C. Hudson, 'Cairngorm snow-field report for 1975', J.M. 1, pp. 284–6; also subsequent years.

16 | Some Notable Snowfalls

Few of us are pessimistic enough (if we live in lowland Britain) to refuse to book business or social engagements between December and March for fear of being stranded in deep snowdrifts. Such things do happen, but are rare enough to go down in history. On the other hand, such is the volume of traffic on our roads that a few centimetres of snow can cause ordinary roads and motorways to become impassable, because of jack-knifed lorries and accidents as the result of skidding. On 5 December 1995 a few centimetres of snow and freezing conditions caused an incredible all-night traffic jam on the M25 between Wisley and Dartford. There was so much traffic that the gritting lorries couldn't get on to the motorway and some people took ten hours to do a journey which normally would take forty minutes. That snowfall was notable, not from the meteorological point of view, but only for the chaos it caused. Some of the more notable snowfalls are described in this chapter.

18–21 January 1881[1]

This is among the greatest snowstorms on record in southern England (Fig. 113). Most of the south had more than 15 cm (6 in); 30 cm (12 in) lay over southern England from Sussex to Cornwall. In Brighton, the average depth was 45 cm (18 in). In Exeter, there were 30 to 35 cm (12 to 14 in), in Kingsbridge, Devon, 45 cm (18 in), in Cullompton 55 cm (22 in) and in Sherborne, Dorset, 60 cm (24 in). A depression had moved from the Bay of Biscay to the Isle of Wight, turned south to France and then moved north-east. England had thus been in the cold sector of the depression all the time. As much as 120 cm (4 feet) fell on Dartmoor.

4–8 December 1882

The storm that passed over northern England and southern Scotland has come to be known as the Border Blizzard. It gave about a metre of undrifted snow. But most of it fell in thinly populated country.

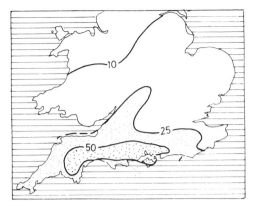

Fig. 113.
Snow depths (in cm) after the blizzard of 18–21 January 1881

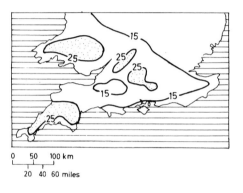

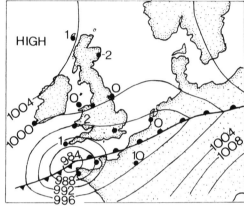

Fig. 114. *(above)* **Snow depths (in cm) after the blizzard of 9–13 March 1891**

Fig. 115. *(right)* **Synoptic situation for 7 h, 10 March 1891. The Blizzard of '91.**

11 July 1888

A light fall of snow occurred with a bleak northerly wind. The fells above Keswick were covered down to 300 metres (1000 feet) and snow was reported in London and the Isle of Wight. In London, the temperature varied between 12° and 6°C (54° and 43°F), and it seems more likely that it was soft hail which fell, at least in the lowlands.

9–13 March 1891: a great blizzard in the south-west.[2]

In Devon and Cornwall the average depth was 60 cm (24 in) with immense drifts (Fig. 114). There were numerous stories of people climbing out of upstairs windows.

The express train that left Paddington at 15 h on Monday, 9 March, got as far as Brent on a southern spur of Dartmoor, where it ran into a huge drift and was imprisoned for four days, eventually reaching Plymouth at 20.30 h on Friday, 13 March. Streets in Bude were blocked and snow was several centimetres deep in the Scillies. On Dartmoor, 120 to 150 cm (4 to 5 feet) of snow fell. The track of the depression which brought this spectacular storm was from the Bay of Biscay north-eastwards along the English Channel (Fig. 115). Westerly winds brought a thaw on the 14th.

18 May 1891

In parts of the Midlands and East Anglia, snow fell to a depth of 15 cm (6 in).

February 1895

A heavy snowstorm in the first half of the month in south-west Scotland, north-west England and the Isle of Man brought traffic to a standstill. The depth was 75 cm (30 in) in some places.

December 1906

A heavy snowfall was general in Christmas week and conditions were very severe in eastern Scotland. It was a white Christmas in London. The snow probably inspired Robert Bridges to compose his poem 'London Snow'. Here are the first two lines:

> When men were all asleep the snow came flying,
> In large white flakes falling on the city brown...

23–26 April 1908

On the 23rd, 30 cm (12 in) lay in parts of Norfolk and Suffolk. On the 24th, 15 cm (6 in) lay at Epsom. On the 25th, Southampton reported 35 cm (14 in), and Totland Bay in the Isle of Wight 25 cm (10 in). On the 26th, there were 27 cm at Marlborough and 45 cm (18 in) at Oxford.

March 1909

In the opening days of this month, Kent and other parts of the south-east had heavy snow, approaching 60 cm (2 feet) in places; 20 cm (8 in) was reported from Walthamstow, north London.

March 1916

A very snowy month in the Pennines. At Wear Head, Durham, it was reported that a continuous blizzard prevailed right through the month. Something like 3 metres (10 feet) of snow fell in the Pennines. The High Peak Railway in Derbyshire was unusable for some time.

16 January 1917

About 12 cm (5 in) of snow fell at Newquay, normally one of the most snow-free parts of the country.

May 1923

In this bleak month a snowstorm raged almost continuously in the Cairngorms, where conditions at 1220 metres (4000 feet) were reminiscent of Spitsbergen at sea-level. In those days there were no skiers or ski-lifts to take advantage of the prolongation of the winter season.

25–27 December 1927 (Fig. 116)

A depression from the Atlantic moved from Ireland to the English Channel and across France to the Mediterranean. It caused a great snowstorm in southern England.[3] About 18 h on Christmas Day, rain in the south turned to snow so heavy that roads were hopelessly blocked by midnight, and a train was snowbound between Alton and Winchester. Even in London, travel was difficult. The storm raged all the 26th with 15 cm (6 in) of snow falling in central London. On Salisbury Plain, the drifts were 6 metres (20 feet) deep. A very impressive sight was clouds of snow blowing out to sea from the cliffs of Devon, Dorset and the Isle of Wight, which seamen on passing ships thought to be fog.

16 February 1929

A small tract on the south-east fringe of Dartmoor to the west of Holne Chase had 2 metres (6 feet) of snow in fifteen hours without any drifting. It was like a cloudburst of snow. Eyewitnesses describe it as coming down as if it were 'shovelled'. This was probably the deepest fall of snow ever measured in a single day's storm in the British Isles at as low an elevation as 300 metres (1000 feet).

23–26 February 1933

A depression moved southwards across Ireland drawing in supplies of moist Atlantic air, while maintaining a bitterly cold easterly flow over most of Britain. A great wet snowstorm ensued as the cold air undercut the milder air. At Harrogate, 75 cm (30 in) fell in three days, and roads over the Pennines were impassable. At Huddersfield, the depth was 75 cm (30 in), and at Buxton 72 cm (28 in). The train,

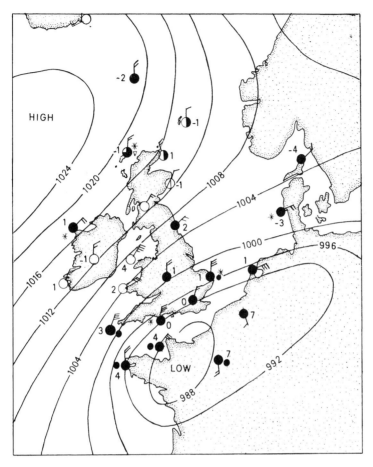

Fig. 116. **7 h, 26 December 1927, the Christmas Blizzard.** Alton, Hampshire, was cut off for several days by 5-metre (15-foot) drifts. Near Royston, Herts, the Baldock road had a 2.5-metre (8.2-foot) drift, 3 kilometres (2 miles) long. East Kent and much of the eastern half of the South Downs had rain.

Royal Scot, arrived at Euston over twelve hours late. At Whipsnade, over 60 cm (24 in) of snow fell. Yet central London, only 50 kilometres (30 miles) away, had a mere 5 cm (2 in), while rain fell on the Sussex coast.

February 1934

At the end of the month, Hull and Holderness had between 30 and 60 cm (1 and 2 feet) of snow, though it soon thawed.

12–19 May 1935

Northerly winds brought cold weather with hail, sleet and snow-showers to many places, particularly on the 17th. There were 12 cm (5 in) at Hampstead, and 11 cm (4 in) at Cockle Park, Northumberland, and at Tiverton, Devon. At West Kirby, Merseyside, snow fell from 9 h to 24.30 h, accumulating to 5 cm (2 in), with the thermometer staying just below freezing point, though it rose above freezing as soon as the snow ceased.

Snowstorm of 26 January 1940

Much of Yorkshire, Derbyshire and Cheshire received between 30 and 60 cm (1 and 2 feet) of undrifted snow and considerable tracts of more than 60 cm. In Sheffield, there was 120 cm (4 feet) of snow in the gardens, and pavements were made impassable for weeks by the vast masses of ice and hard compacted snow. Liverpool, Southport and the coastal plain of Lancashire, usually free of serious snowfalls, had a downfall as heavy as in Yorkshire. An express train was buried in a drift only a few kilometres from Preston station. This great snowstorm extended to Scotland, where the West Highland line was completely blocked, a 6-kilometre stretch being covered by from 1 to 3 metres (3 to 10 feet) of snow. Conditions on the higher Chiltern ridges were only a little less severe than those further north. In London, the heavy drifting snow was remarkable. Other parts of the country were affected by the ice-storm described in Chapter 17. On the 27th, 25 cm (10 in) of snow lay at Eastbourne. On the 28th, there were 37 cm (15 in) at Pontefract and 60 cm (2 feet) at Malvern. Bird life suffered severely and some of the finest herds of deer perished in Scotland. On Exmoor, there were drifts between 2 and 2.5 metres (6.6 and 8 feet) deep and it was possible to walk over hedges and gates without realising it.

1941

January and February 1941 were also very snowy. There were 37 cm (15 in) of snow at Birmingham. Hoylake in the Wirral had 30 cm (12 in) and at Hillsborough, Co. Down, there were drifts between 2 and 4 metres (6 and 13 feet). Parts of Caithness and Sutherland were isolated for several days. At Balmoral, the equivalent of 50 cm (20 in) of undrifted snow fell. The duration of the snow-cover was not as long as in 1940. Snow lay throughout January at Balmoral; also at Forres, where it continued throughout February. There were twenty-two mornings of snow-cover in Edinburgh, twenty-three at Birmingham and even eleven at Bude. Snow lay on the ground at Durham on only two mornings at an altitude of 102 metres (335 feet), but on twenty-five at Ushaw College (Co. Durham) at 181 metres (594 feet). In the London area, snow lay on twelve mornings at Hampstead and five at Kensington. On 26–27 March, between 75 and 90 cm (30 to 35 in) fell in north-east Scotland around Tain, and even 22 cm (9 in) at Oban on the 30th.

1942

In the third winter of the war, in January and February, the eastern half of the country was notably more snowy than the west as several depressions moved south-east across England. There was a general snowfall of about 15 cm (6 in) in the Midlands and the south-east in early February. Near Canterbury there was snow-cover on forty-three mornings in this winter. For the third successive year north-east Scotland was badly affected. Planes had to be used to drop food to passengers in a stranded train near Forsinard.

8 May 1943

A northerly airstream brought thick snow to virtually all Scotland and northern England, covering even the small Hebridean islands, and giving a 7 cm (3 in) fall at Duntulm, Isle of Skye, and 15 cm (6 in) on the Isle of Man.

26–27 February 1944

A depression moving north-east brought very cold weather to Scotland and Northern Ireland, with much snow. There was severe drifting in the Isle of Skye and in Caithness. In the northern half of England, there were 30 cm (12 in) of snow at Mansfield, Belper and Nottingham, and a few centimetres less in the Border Country.

8–12 December 1944

Roads were blocked by deep drifts near Ullapool, west Scotland.

25–29 January 1945

A relatively narrow band of country from South Wales to Yorkshire had a very heavy snowfall; 100 cm (40 in) of undrifted snow accumulated at Penarth, near Cardiff. On the promenade exposed to the wind, drifts 2·5 metres (8 feet) high piled up.[4] Eyewitness accounts say the snow was powdery and most fell in three hours during the night. Unfortunately, there was a blackout on reporting any weather news in case the enemy might gain some advantage from the information. Enquiries in South Wales have so far failed to produce photographic evidence. For several days, road and rail traffic was totally disrupted. Had such a fall occurred today there is no doubt that it would have provided a feast for the media, as the 1978 blizzards did in the south-west and the Thames estuary storm of 1987.

On one day the sun shone on to the unusually smooth waters of the Bristol Channel and the reflections of the snow-covered Somerset hills in the sea were a magnificent spectacle. On the evening of the 29th, further snow turned to rain about midnight and a rapid thaw followed.

Elsewhere during this snowy month, there were drifts of 6 metres (20 feet) in Morayshire, and between 3 and 6 metres (10 and 20 feet) on the Glasgow to Argyll road. Over 60 cm (2 feet) of snow fell in Northumberland on the 9th. There was a fall of 25 cm (10 in) in Edinburgh on the 29th before the thaw came. Few parts of the country escaped the snow, which lay on the ground on twenty-seven mornings at Inverness, twenty-one at Armagh, thirteen at Tiree, fourteen at Kew, fifteen at Worthing, ten at Newport, Isle of Wight, and Cardiff, and eight at Newquay, though neither Penzance nor Scilly had a single morning's snow-cover.

The winter of 1946–47

There were so many snowfalls that it is not possible to mention them all. There were significant falls in December, and on 4 and 5 January, followed by a very mild interlude. But this notoriously hard winter did not really start until the third week of January. By the 30th, snowstorms had given several centimetres to the Scilly Isles and the Lizard, with a general covering of between 7 and 30 cm (3 and 12 in) over southern England. Deep drifts piled up in the south-west on the higher ground as a slow-moving depression transferred from the Channel westwards, then moved south-east into France. Over Exmoor, there were three violent blizzards, on 28 January, 23 February and 4–5 March, when snow fell without pause for thirty-six hours.[5] Snow-cover was continuous over low ground away from the west coast from the third week of January and throughout February (Fig. 117). There were 30 cm (12 in) of snow or more in Co. Durham, Yorkshire, Lincolnshire, and the East Midlands by 5 February. At Ushaw College (Co. Durham), snow was 37 cm (15 in) deep on the 5th, 55 cm (22 in) on the 18th, and 70 cm (28 in) on the 28th. At Huddersfield, depths varied between 50 and 37 cm (20 and 15 in). At Waddington, Lincolnshire, the greatest depth was 37 cm (15 in) on 22 February. In western Scotland there was very little snow. Pressure was high over Scandinavia and Greenland, and depressions moved on more southerly tracks than usual (track G, page 30).

Heavy snow occurred on 4 and 5 March when a deep depression approached the south-west and moved along the Channel (though it had earlier showed signs of moving across southern England), bringing Tropical Maritime air and a thaw in its wake. In fact, the snowstorm which followed was one of the worst of the winter in some parts, causing severe disruption to Southern Railway trains; a journey from London to Brighton taking eight hours. Further north at Lake Vyrnwy, snow lay 90 cm (35 in) deep. At Crickhowell, there were 55 cm (22 in), and at Wrexham 53 cm (21 in). On 10 March, snow was 40 cm (16 in) deep at Birmingham, and a couple of centimetres deeper at Durham and Harrogate. In parts of eastern Scotland there were immense drifts up to 8 metres (26 feet) deep, isolating farms and villages.

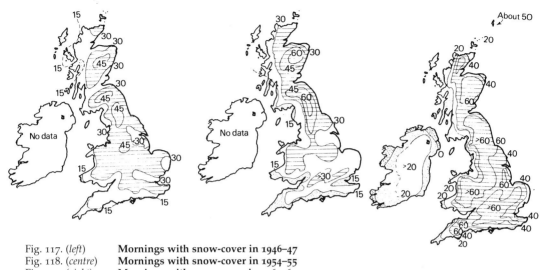

Fig. 117. (*left*)   **Mornings with snow-cover in 1946–47**
Fig. 118. (*centre*)   **Mornings with snow-cover in 1954–55**
Fig. 119. (*right*)   **Mornings with snow cover in 1962–63**

20–23 February 1948

Deep powder snow fell over much of south-east England. There was snow on the ground as far west as Cornwall, in an intensely cold spell in an otherwise very mild winter. The snow-cover was not removed by a thaw with rain but by bright sunshine which gradually melted the snow-cover and by sublimation (evaporation) in the shade.

24–27 April 1950

A depression moved south over the North Sea and brought very cold Arctic air in its wake. There were widespread snowfalls which affected the south and south-east most severely.[6] Many trees already in leaf were badly damaged and telephone wires were brought down by the weight of the snow which contained a lot of water, providing an excellent example of a wet and damaging snowfall. In London, there was a complete covering of snow on the pavements on the morning of the 26th, a rare enough event in midwinter. In a broad belt from Kent to Wiltshire there were about 13 cm (5 in).

December 1950

Incursions of Arctic air provided several snowy spells. On the 4th, there were 50 cm (20 in) at Dalwhinnie and 15 cm (6 in) at Manchester. Later in the month there was a 25 cm (10 in) fall at Bournemouth, 37 cm (15 in) in the Isle of Wight, 35 cm (14 in) at Scarborough and Lowestoft, and 23 cm (9 in) at Southport. Generally, the snow-cover was not prolonged, though the snowfalls were mainly of dry snow.

The winter of 1950–51

This was probably the worst winter of the century on our moors and mountains. At Braemar there was snow-cover on 103 mornings, many more than in the notorious winter of 1947. Polar Maritime air was frequent during the winter. At low levels, rain, sleet or wet snow was common, but with a high lapse-rate only snow occurred on the higher ground. Depressions which passed over or near the British Isles moved quickly so that the warm-sector air did not have much time to thaw lying snow before colder air and more snow arrived. Cross Fell was not clear of snow until 10 July.[7]

January 1952

A very snowy month in Scotland and Ulster. Snow fell on nineteen days at Eskdalemuir and seventeen at Aldergrove, near Belfast.

28–30 April 1952

The northern suburbs of London had 15 cm (6 in) of snow. Over the Chilterns and Cotswolds there were 30 cm (12 in).

26–30 November 1952

25 cm (10 in) fell at Whipsnade in the Chilterns and 20 cm (8 in) in the Outer Hebrides. The Wirral had a 30 cm (12 in) fall from a northerly airstream.

29 January–7 February 1954

There was deep snow in Hampshire and Dorset and as far east as Haslemere and Hindhead, where snow fell all day and was over 30 cm (12 in) deep by evening. Not 20 kilometres (12 miles) away in Guildford, no snow fell at all as the frontal cloud did not extend that far east. The depression which brought the snow moved from Cornwall to France, leaving the east in cold Polar Continental air.

January 1955 (Fig. 118)

On 3–4 January snow spread southwards over England, giving a fall of from 10 to 30 cm (4 to 12 in) with deep drifts over the moors. After a milder spell, a cold front on the 16th, preceded by almost total darkness in central London around noon and over the south coast during the afternoon, brought a sharp drop in temperature and further snow. Depths were not outstanding in the south, 15 cm (6 in) in London, but in north-east England and Scotland the winter was snowier than 1947. Aberdeen and Caithness had 60 cm (2 feet) of undrifted snow; farms and outlying villages had to be supplied by plane, in 'Operation Snowdrop'. Two passenger trains were snowbound near Wick, while even Blackpool had snowdrifts of 1·5 metres (5 feet) on 19 January.

May 1955

High-level roads across the Pennines and Peak District and in South Wales were blocked by snow, and snow-ploughs were needed before the flow of traffic could be resumed. Snow fell all day in Sheffield and on Salisbury Plain, even reaching parts of the south coast. At Malham Tarn Field Centre in the Pennines, snow fell on eight days and 10 cm (4 in) was lying on the 17th.

February 1956 [7]

High pressure from Russia to Scandinavia produced intense dry cold and kept depressions at bay during most of the month. There was consequently very little frontal snowfall from layer clouds, and snow of any depth was confined to the east coast from Yorkshire to Kent. Here, villages were cut off by deep drifts, though not many kilometres away snow-cover was negligible. Snow fell on eighteen days at Heathrow, mainly in the form of flurries from thin low clouds. There was barely enough to cover the ground on three mornings, even if temperatures were, for the greater part of the month, well below freezing-point. The snowfalls on the hills in Kent and Yorkshire provided a good example of instability and orographic uplift affecting the distribution of lying snow. Frontal snow affected west Cornwall, giving a fall of 25 cm (10 in), but that did not lie for long.

1961–62

Deep snow covered London and much of the south-east on New Year's Eve. There were 37 cm (15 in) of lying snow on 1 January over the Chilterns, and between 30 and 35 cm (12 and 14 in) over parts of the Midlands. The snow remained for about a week in the south-east. There was further snow at the end of February, and 25 cm (10 in) accumulated at Jersey.

16–20 November 1962

Cold northerly winds brought snow and sleet to most parts except the extreme south-west. Level snow was 30 cm (12 in) deep in parts of Scotland. At Tredegar, South Wales, there were drifts of 3 metres (10 feet) on the 18th. Little Rissington in the Cotswolds reported seven mornings with lying snow in November.

1962–63

On 23 December, a ridge extended from the southern Baltic to south-west England allowing very cold Polar Continental air to spread across southern Britain. Meanwhile, a trough of low pressure moving south brought rain to northern Scotland on Christmas Eve, turning to snow as it moved southwards. On Christmas morning, level snow in Glasgow was 7 cm (3 in) deep. On Boxing Day morning, after a hard frost, snow began to fall thickly over much of England. By evening, many places had over 30 cm of snow (12 in), and driving was virtually impossible even in lowland West Sussex. Behind the trough, conditions were still very cold, even though Arctic air had replaced continental air.

On the 29th, a trough moved slowly north to the Channel and brought a severe blizzard to the south, with fine powdery snow forming immense drifts in gale-force easterly winds. The falling and drifting snow was a sight not to be forgotten in the light of the street lamps, even in Guildford. Many towns and villages were completely isolated by snowdrifts. Even the main Guildford to Leatherhead road was impassable near Horsley for a short time. Some narrow sunken lanes near Peaslake, Surrey, remained completely blocked until March. Driving was quite out of the question on Dartmoor and Exmoor.

At the beginning of January, freezing rain and drizzle made conditions treacherous. Often freezing rain is followed by a thaw. But this time the stronger cold air pushed back the mild weather. Further heavy snowfalls followed in the south and south-west as the warm front retreated, isolating many places for the second time in a week. Only light additions of snow occurred during the next fortnight. On 19 January snow and gales spread across the country, followed once more by freezing rain, which fell on top of lying snow. Snow-depths at the beginning and end of January are given in Table 31.

Table 31 Snow Depths in January 1963

| | 1 January | | 31 January | |
	cm	in	cm	in
Hampstead (London)	30	12	22·5	9
East Grinstead (E. Sussex)	75	30	15	6
Spadedam (Cumbria)	37	15	62·5	25
Tredegar (Gwent)	80	32	100	40
Princetown (Devon)	45	18	50	20

On 4 February, a depression became slow-moving to the west of Ireland. There was level snow in excess of 50 cm (20 in) in Belfast. There were then periods of snow which in some areas produced the worst blizzards of the winter, with snow falling continuously for thirty-two hours. Farms and villages in the moors were isolated again, some for the tenth time since Christmas. On the 15th, there was heavy snow in Cumbria and the Borders. On the 22nd and 23rd, several centimetres of snow fell in Surrey and London, hardly worthy of mention apart from the freshening effect of new snow on top of old dirty snow. It was possible to estimate the number of snowfalls by counting the dirt lines in a snow section. Unlike the great snows of 1947 and 1945 when heavy rain came with the thaw, this snowy winter ended with a spell of beautiful sunshine, so that there was little snow left to wash away when the rains came on

6 March, and there was no flooding. Table 32 shows mornings with snow-cover in this winter.[8] The data are mapped in Fig. 119.

March 1965

There were drifts up to 3 metres (10 feet) high on Salisbury Plain from a general fall of 22 cm (9 in). The Peak District had a heavy fall on the 22nd, with 35 cm (14 in) falling at Buxton.

November 1965

Snow covered the ground at some south coast resorts, and Scarborough had snow on the ground on seven mornings.

1966

There was general snow in eastern England from 14 to 22 January, with 17 cm (7 in) at Ushaw, 10 cm (4 in) at Durham, 25 cm (10 in) at Dover, and 20 cm (8 in) on Alderney. On 15–16 April, there was widespread snow. Unusually for April, the snow was of a powdery nature in many places as the afternoon temperature was 0°C. About 15 cm (6 in) accumulated on the North Downs during the day. Pressure was high to the north-east and low to the south-west. During the month, snow-cover was reported on two mornings at Hampstead and Wisley, three at Tiree and eight at Huddersfield.

December 1967

Snow was 43 cm (17 in) deep in central and North Wales on the 9th. There were 15 to 23 cm (6 to 9 in) in the West Midlands and mid-Devon. Two small polar lows moved south-south-east from Scotland, the first giving about four hours of snow to most places from North Wales to mid-Sussex. Over the south coast something more unusual happened. As the very cold air came into contact with the relatively warm sea, intense convection caused an outbreak of thunderstorms along the Channel and there was eight hours' snow in Brighton (see Fig. 120). The area of heavy snowfall was restricted to the shore and promenade where there were 28 cm (11 in). The snow-cover remained until milder air arrived on the 11th.

February 1968

At Buxton there was snow cover from the 3rd to the 28th. On the 6th, between 30 and 45 cm of snow (12 to 18 in) fell around Birmingham and Manchester, bringing traffic to a standstill.

Table 32 **Mornings with Snow Lying in the Winter of 1962–63**

	November	December	January	February	March	TOTAL
SCOTLAND						
Lerwick	8	6	19	19	0	52
Stornoway	3	3	3	0	0	9
Tiree	0	0	0	1	0	1
Braemar	8	13	31	28	6	86
Edinburgh	1	5	25	25	0	56
Eskdalemuir	9	9	31	28	6	83
ENGLAND						
Cockle Park	5	6	28	28	6	73
York	5	4	28	28	5	70
Gorleston	0	6	8	14	0	28
Cambridge	0	5	31	28	2	66
Birmingham	7	7	31	28	2	75
London (Regent's Park)	0	4	31	23	0	58
Folkestone	0	5	31	12	0	48
Shanklin (Isle of Wight)	0	3	5	4	0	12
Plymouth	4	6	5	0	0	15
Totnes (South Devon)	0	4	31	5	0	40
Falmouth	0	0	3	4	0	7
Scilly	0	0	0	1	0	1
Ambleside (Cumbria)	2	3	3	27	0	35
WALES						
Llandudno	0	1	8	6	0	15
Aberystwyth	0	0	20	8	0	28
Cardiff	0	6	29	16	0	51
NORTHERN IRELAND						
Kilkeel (Co. Down)	0	0	0	0	0	0
Armagh	2	5	27	4	0	38
REPUBLIC OF IRELAND						
Dublin Airport	0	2	6	0	0	8
Cork	0	2	28	0	0	30
Shannon	0	1	9	0	0	10
Valentia	0	0	0	0	0	0
CHANNEL ISLES						
Guernsey	0	7	3	7	0	17

December 1968

There was a white Christmas in many places as a depression moved east-south-east across Cornwall. Rain turned to snow with 10–15 cm (4–6 in) across southern England. Further north, Whitby was cut off by blizzards. There were 40 cm (16 in) of snow in Nottingham and 50 cm (20 in) at East Dereham, Norfolk.

February 1969

On the 19th, a severe easterly gale accompanied by snow gave 25 cm (10 in) in five hours in Derbyshire. There were 30 cm (12 in) at Deal, blocking roads and isolating villages. In Northern Ireland there was snow lying on fifteen mornings. Fig. 121 shows a fall of dry snow around St. Paul's Cathedral. Much of this month's snow came with northerlies and, as Fig. 110 shows, central southern England had virtually none at all.

25 December 1970 (Fig. 122)

This was London's seventh white Christmas of the century. Substantial falls had occurred at Christmas in 1906, 1927 and 1938, and slight falls in 1917, 1923 and 1956. High pressure to the south of Iceland with a ridge extending to Norway brought cold weather southwards. Snow fell late on Christmas Eve and in the early hours of the 25th to give a white Christmas to the London area and the south-east. There were 20 cm (8 in) towards the coast and over the hills, about half this amount falling generally over low ground. Many places outside the towns had snow-cover for the Twelve Days of Christmas. The western edge of the snow lay between Farnham and Basingstoke. In the West Country, snow fell on Boxing Day.

October 1974

Northerly winds brought severe gales and snow. On the 28th, the Perth to Braemar road was temporarily blocked. Snow also fell in Kent but did not settle. Yet it brought forebodings of a hard winter at a time when fuel-oil supplies were severely restricted by the Organisation of Petroleum Exporting Countries (OPEC).

March and April 1975

Easter was much colder than Christmas and there was much snow, especially over the hills. On 8 April, snow fell all day at Aberdeen. Glenmore Lodge had 45 cm (18 in) of snow. In Kent, 15 cm (6 in) was lying on the morning of the 9th. In most instances the snow-cover was not continuous but was renewed during the night after a general or partial daytime thaw. In the south, this snow was the first that many children under eight years old had seen.

Fig. 120. **Rottingdean, near Brighton, Sussex, 8 December 1967**: a convoy of cars, buses and lorries going nowhere after a heavy snowfall (Topham Picturepoint).

Fig. 121. **February 1969.** A rare sight in London, as a northerly airstream brings a blanket of powdery snow to the streets around St. Paul's Cathedral. Another powdery snowfall around St. Paul's is immortalised by Robert Bridges in his poem 'London Snow' (see page 189).

June 1975 [9]

'Rain stops play' is a familiar situation in cricket matches. Snow has been known to stop play in April or May. But when before had this happened in June? On the 2nd, snow prevented play at Buxton and *The Times* carried a photograph of the snow-covered cricket ground and a few disconsolate spectators. Play was also interrupted at Colchester. John Arlott, the well-known cricket commentator, also saw snow at Lord's on 2 June. *The Daily Telegraph* of 24 October 1975 had the following passage: 'A complaint that John Arlott falsely stated in *The Guardian* that snow fell at Lord's on June 2nd, and that the newspaper refused to print a correction or a letter pointing out the absurdity of the report, has not been upheld by the Press Council.' Apparently a reader of *The Guardian* did not believe John Arlott had seen snow and complained to the Press Council about the report.

There were many accounts of snow on 2 June. In the early afternoon snow began

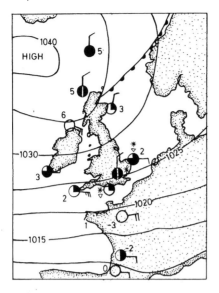

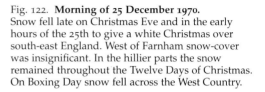

Fig. 122. **Morning of 25 December 1970.**
Snow fell late on Christmas Eve and in the early hours of the 25th to give a white Christmas over south-east England. West of Farnham snow-cover was insignificant. In the hillier parts the snow remained throughout the Twelve Days of Christmas. On Boxing Day snow fell across the West Country.

to lie in the centre of Edinburgh: it could already be seen lying on the Pentland Hills, above 150 metres (500 feet). Snow was observed in Birmingham and in the Newark area, where it covered the ground for a brief period in the early morning. At about 8 h snow also fell at Grantham and Peterborough.

It seems likely that a small depression or polar low moving south in the very cold Arctic air was responsible for the unseasonable weather. Other reliable observations of sleet or snow falling in the London area in summer are to be found for 21 May 1821 and Derby Day, 22 May 1867. There is said to have been snow on the Downs in Sussex on 12 June 1791, but the distinction between snow and hail is not always appreciated, which is not surprising when events like the Tunbridge Wells hailstorm occur (Chapter 8).

The heavy wet snowfall of 14 January 1977 (Fig. 123)

On 13 January, a depression formed out in the Atlantic, deepened rapidly and moved across the Midlands. Heavy snow fell, commencing around mid-morning in the Home Counties. Between 5 and 10 cm of slush accumulated on roads outside London, delaying traffic. In the late afternoon steady rain began to fall as the warm front moved northwards. By the next morning there was virtually no lying snow in places to the south of the depression track, but farther north the snow-cover lasted a week. Throughout the snowfall (in Surrey) the temperature remained fractionally above freezing-point, and with the air saturated there was no evaporative cooling and water

Fig. 123. *(facing page)* **The heavy wet snowfall of 14 January 1977**

Norfolk

Ireland

S N O W

R A I N

dripped copiously from trees and wires, while gutters became choked with slush. The satellite photograph shows the cloud pattern at 18·46 h. Frontal cloud had not reached Norfolk, nor much of northern England. Parts of western Ireland can also be seen.

6 June 1977, Silver Jubilee Day

Those who decided to go and see the beacon on Skiddaw had a wintry experience for there was snow there. The next morning the snow could be seen from the valley below.[10]

27–29 January 1978 [11]

On the 28th, a depression moved slowly east across the Midlands so that cold north and north-east winds covered Scotland. Fig. 124 shows the synoptic situation at 10 h on this day. The dense frontal cloud responsible for the heavy snow can be seen clearly in the satellite photograph (Fig. 126). Transport was badly disrupted, cars were marooned and three people lost their lives when they were buried alive. The story of one man will long be remembered. He survived burial in the snow for eighty hours. Some stocking samples he was carrying in his car had kept him warm. He had wrapped them round and round himself. It is estimated that 70 cm (28 in) of level snow fell. At a village a few kilometres from Cape Wrath a reporter from *The Times* was told: 'The worst problem is the deep-freeze. Without power, the contents will soon start to defrost, and in a place like this we have to live from the deep freeze. I might lose £150 of food unless the power comes back soon.' Fortunately, milder weather arrived by the 31st, and power and transport services were restored. These occasional severe storms remind us how dependent our way of life is on electric power.

February 1978 [12]

This was a very snowy month in the north-east, east and south-west. In Devon, Somerset, Dorset and parts of South Wales, February 1978 will be remembered for a long time to come. A broad band of country from the Thames Valley to Merseyside had no significant snow at all. The cold weather began on the 8th, as a high over northern Scandinavia fed Polar Continental air to Britain. By the 10th, snow showers were widespread and as much as 10 cm (4 in) fell in an hour at Middlesbrough. Tunbridge Wells had 13 cm (5 in). On the 11th, snow was lying in the Scilly Isles for the first time since 1963; both Edinburgh and Durham reported 30 cm. The Home Counties experienced only snow flurries. On the 15th and 16th heavy snow affected Plymouth where snow ploughs had to be used. On the 17th, 45 cm (18in) of snow fell in Dorset. Two days later, an intense depression deepened off the south-west and a severe blizzard followed as the front separating very cold continental air from warm tropical air (with temperatures up to 20°C (68°F) over south-west France) became

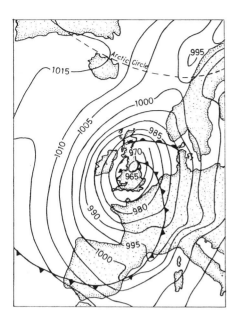

Fig. 124.
**The Highland Blizzard,
noon, 28 January 1978**

Fig. 125. **Blizzard in the South-West,
noon, 18 February 1978.** The synoptic
situation just before the climax of the
blizzard. Six hours later, many towns
and villages in Devon were cut off by
drifts between 3 and 8 metres (10 and
26 feet) high. By the 22nd, milder air
from the Atlantic had replaced the cold
easterlies and there was a general thaw.
Without a doubt this slow-moving front
produced one of Britain's memorable
blizzards comparable with those of
1963, 1947, 1891 and 1881. One major
difference between 1891 and 1978 was
that Cornwall escaped this time, being
in the warmer air.

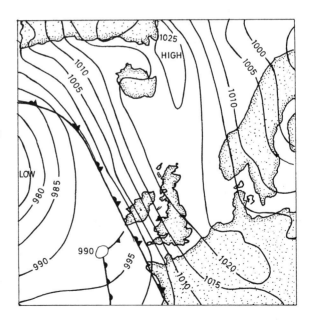

Fig. 126. (*facing page*) **The Highland Blizzard, 9 h, 28 January 1978**. A deep depression moved across Scotland. Weather was quite mild in the south of England, but a severe blizzard developed in northern Scotland. There were deep drifts on low ground in Caithness and Sutherland (Highland Region). As much as 70 cm (28 in) of snow fell over the Grampians. The northern extent of the uplifted tropical air is shown by the arc of upper cloud.

slow-moving (Figs. 125 and 127). Enormous drifts piled up in the gale-force winds from the Mendips and Dorset in the east to Dartmoor, Barnstaple and South Wales in the west. Snowfall extended along the South Downs and north-west into Wiltshire, across the Cotswolds, and into South Wales. Farnham, Swindon and Gloucester had no snow at all from the 'South-West Blizzard'.

On Dartmoor, 18 February 1978, an eyewitness account [13]

'The weather forecasters were predicting heavy snow with gales for Saturday, 18 February. Derek asked me how I would like to experience the start of a blizzard on top of Dartmoor. The idea was appealing and exciting, but caution governed my response. "Let's wait and see what tomorrow, Saturday, brings and let's pay heed to the morning's forecast" was my considered response.

'Saturday dawned grey and cloudy with a rising easterly wind. Overnight the screen temperature had fallen to −7°C, but no further snow had fallen. The forecast was for gales and heavy snow with blizzard conditions over the high moors of south-west England.

'We breakfasted in the warm Devonshire kitchen. " How about it, then", enquired Derek. "Game to risk your life?" I laughed. "I'm game, fortune favours the brave." In truth I thought it foolhardy, but the once-in-a-lifetime chance of high drama on Dartmoor could not be lightly dismissed. Besides Derek's wife knew where we would be, and we could exactly pinpoint our position on the O.S. map.

'We parked the car at our selected spot at 11.47 GMT on the B3212 (the Tavistock-Moretonhampstead road). Flurries of light snow were falling but the views nevertheless were quite magical.

'We set off, booted and muffled up against a force 8 easterly gale. But it was not daunting, indeed it was exhilarating, so that early fears evaporated in our receding breaths. The whipped-up snow from previous falls cut our faces, but despite the vanguard snow flurries I could see a watery sun, and when the drifting snow allowed a horizon line. "I'm glad I came", I shouted, "it's terrific." We had gotten over half way (of the planned three miles) and I felt good, and with rising confidence that we would make it OK.

'It was 12.30 and we had made splendid progress despite the head wind and the knee-high drifts in places. I looked back. "Here it comes," I bawled. To the west the horizon was blotted out.

'I was totally unprepared for the next minutes. Nothing I had read in literature had prepared me. Neither tales from the Arctic, the Cairngorms nor Dartmoor itself. In an instant we were enveloped in a freezing white hell and a screaming wind that tore

at us in all directions. We were literally stopped in our tracks, unable to think or walk or speak or see. In an instant we were blinded, struck dumb by elemental forces beyond imagination. The real terror was the inability to breathe. I was sucking snow into my lungs and my nostrils were furred up with encrusting snow... But the Gods were with us. A lull came in the wind, although the atmosphere was still thick with choking powdery snow, and shapes could be discerned, and within a stone's throw was the blessed lump that could only be the car... then came the awful realisation that the lock was hopelessly frozen. It was Derek's turn to try the lock. He silently took the key from me and put it in his mouth over the door lock with the key jutting out from the side of his mouth... Even inside the car the wind had forced snow through.'

Fig. 127. *(facing page)* **The day before the Great South-West Blizzard,** February 1978, in Devon, Somerset, Dorset and South Wales. Pressure is low in mid-Atlantic and high to the north of Iceland. Very cold Polar Continental air is sweeping across southern England. Frontal cloud covers Cornwall where some snow had already fallen. This cloud marks the interaction of warm moist Tropical Maritime air over France, with a temperature around 20°C (68°F) over south-west France and continental air from Europe, well below freezing-point. An indication of snow cover before the blizzard can be seen. From Suffolk northwards to Scotland, there was a general layer of snow which had persisted since the end of January in some parts. There were also several centimetres of snow over the hillier parts of the south-east. The most populous part of the country, from Surrey across to Merseyside, had no snow on the ground. All this region escaped any substantial snowfall during the whole winter as it was totally unaffected by the blizzard, and sheltered from snow showers on east and north-east winds. The cloud pattern in the sea, from Lincolnshire northwards, indicates convective activity caused by rapid warming of the surface layers of the very cold air by the relatively warm sea. Because cold air can hold so little water vapour when fully saturated, the clouds are thin and showers will not penetrate far inland, which partly accounts for the snow-free zone across the centre of England. The snow was actually forecast to spread right across the south; but the front unexpectedly slowed down and deposited its load in the south-west. The sky was very leaden in Guildford that day and the ground was frozen solid, so that muddy farmyards could be negotiated with ease by walkers.

Fig. 128. **Snow depths (in cm) from the blizzard of 18–20 February 1978**

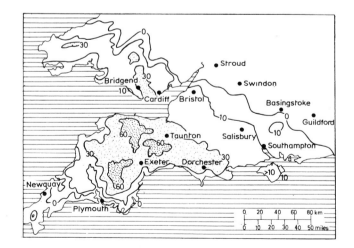

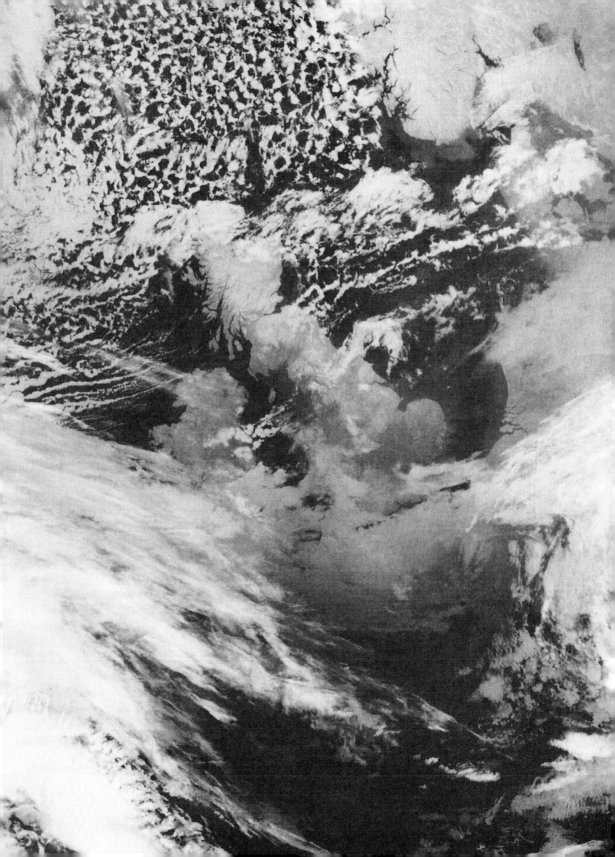

Fig. 129. **Deep snow at Minehead, Somerset, after the blizzard of 18–20 February 1978**

This abridged version of a fascinating description of the famous Dartmoor storm is a useful reminder to those of us who might venture above 500 metres (1600 feet) in wintry weather.

The winter of 1978–79

Snow-cover was continuous through January and February above about 300 metres (1000 feet) over England and Wales away from the west. Around Pontypool, in Gwent, the mountains were snow covered from 30 December to 17 April. Of all the places listed in Table 29 only Buxton had longer snow-cover in 1978–79 than in any other winter back to the 1930s, but at low levels there were occasional thaws.

In January, there was snow-cover on between fifteen and twenty-five mornings away from the south and west coasts. During February, there was snow-cover on only two mornings at Brighton, three at Southport, four at Redcar, from five to ten in the London area, but twenty-eight at Sheffield and Oakham.

Easterly winds brought sleet and snow to the northern half of the country in the week after Christmas. The equivalent of 45 cm (18 in) of snow fell over the Pennines, blocking roads and isolating farms and villages. By the 30th, wintry weather had spread south in association with a depression moving east along the Channel. In the London area, fine powder snow was driven by strong winds; however, the total amount that fell was not large, so there were no impressive drifts in the Home Counties. In Lincolnshire, an unusual combination of wind and snow texture produced 'snow rollers' (Fig. 130). On 1 January, Tiree had 12 cm (5 in) of snow. The following day, Manchester and Stoke-on-Trent had a heavy but localised fall of 20 to 25 cm (8 to 10 in). On the 3rd, south Devon was affected by severe gales and snow, from a depression which moved south-east into France. The A38 was blocked, while drifting snow, blown off the fields, continued to dislocate road traffic. Northern Ireland experienced heavy snow on the 10th, many places having 30 cm (12 in). Sheffield had 36 cm (14 in) on the 19th, and London had 10 cm (4 in) on the 23rd.

Fig. 130. **Snow rollers in Lincolnshire, January 1979.** An unusual combination of wind and snow texture caused these natural snowballs.

In the Peak District,[14] more snow began to fall on the evening of 13 February, continuing the next day in the form of frequent showers accompanied by a force 7 wind. Buxton was cut off and about forty main roads were blocked. At Middleton, Derbyshire, at 207 metres (679 feet), there was a maximum temperature of −4°C (25°F). Holly and laurel bushes that had not been buried by snow lost their leaves, and winter vegetables not covered were killed off by the cold. East Anglia and Kent were badly affected, too. Villages in Norfolk were isolated and even big centres such as Norwich and Great Yarmouth were cut off for two days. After a milder spell, the West Midlands had over 30 cm (12 in) of wet snow from a slow-moving depression on 16 March. The next day Newcastle upon Tyne was virtually cut off, with huge drifts piling up in Northumberland and Durham. As the depression moved slowly north-eastwards, Scotland had snow. On 21 March, there were 25 cm (10 in) in Edinburgh, and 15 cm (6 in) in four hours in Fife.

6 November 1980

There were 5 cm (2 in) of snow in Jersey. It remained on the ground for most of the day. A cold easterly current affected the Channel Isles. This was one of the earliest falls of snow for Jersey.

April 1981

Warm weather in mid-April with temperatures of over 20°C (68°F) was followed between the 20th and 27th by a very cold spell with maxima as low as 1°C at Nottingham. On the 23rd there was prolonged snow over north-east England. At Low Etherly, Co. Durham, 23 cm (9 in) of snow accumulated. Above about 300 metres (1000 feet) there were drifts 5 metres (16 feet) high causing great problems for hill farmers. Sheffield had 11 cm (4 in) of snow on the 24th, followed by a daytime thaw; 26 cm (10 in) of fresh snow on the 25th and a further 11 cm the following morning. Wiltshire and Gloucestershire bore the brunt of gale-force winds and driving wet snow on the 25th. Over 300 people were trapped in cars or coaches on Salisbury Plain overnight and had to be rescued the next morning by the army or police. The combination of high winds and wet snow caused widespread power failures and some towns and villages were without power for forty-eight hours. Outside the affected area there was continual flickering of the electric lights. Snow depths on the morning of the 25th were: 50 cm (20 in) in the Peak District; 35 cm (14 in) over the hills above Leominster; 20 cm (8 in) at Cosby, Leicestershire; 10 cm (4 in) at Hall Green, Birmingham; 8 cm (3 in) at Axbridge, east Devon; 7 cm (3 in) at Trowbridge, Wiltshire. To the east of a line from the Wash to Southampton there was little or no snow. Instead there was heavy rain, particularly over East Anglia where 90 mm (3·5 in) fell in two days. This snowstorm can be compared with the storm in April 1908 (page 189). On that occasion there was more snow on the low ground, whereas in 1981 snow only settled to any extent above about 100 metres.

December 1981

On the morning of the 8th, rain turned to snow in London and many parts of the south in the rush-hour, and snow began to settle even on wet roads almost at once, causing many delays. There were 15 cm (6 in) of snow over the Chilterns and 8 to 10 cm (3 to 4 in) over a wide area, though the Channel coast had none. By mid-afternoon, the sun was shining on a Christmas card scene, as every twig, branch and piece of wire had a thick, tenacious covering of snow: the first of the snow was very wet and, as it froze, it bonded itself tightly. A further snowfall on the 11th, deriving from a depression to the south-west, gave 18 cm (7 in) of snow to Herefordshire and much of the south and East Anglia. At Heathrow, the total snow depth was 26 cm (10 in). These two snowfalls did much damage to trees and shrubs, snapping branches and boughs. Fortunately for vegetation, if not for car drivers, the next snowfall on the 13th was accompanied by gale-force winds, which shook snow off the trees and prevented accumulation. On Salisbury Plain, on the A303 at Wylye, some 35 kilometres (20 miles) from Andover, a number of people were trapped in their vehicles for nearly 18 hours from 2 p.m. on the 13th; for that stretch of road was a hedge-lined single carriageway then. The drifts formed on the sheltered side of the hedge. A group of vehicles had to stop on an incline. Snow was falling and drifting so fast that when the drivers realised what was happening, they could not open their doors, and they had to stay inside until they were rescued. As it happened, a fairly rapid thaw made this easier than it might have been.

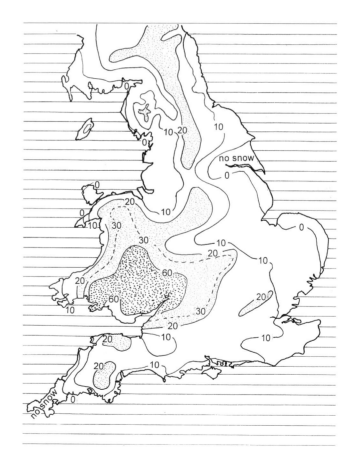

Fig. 131. **Snow depths in January 1982.**
In Gloucestershire and South Wales the snow was more than 60 cm (24 in) deep: enough to stop traffic. Note the unusual lack of snow in the east.

The month will rank as very snowy. In Scotland, there was snow-cover at 9 h on twelve mornings at Edinburgh and on thirteen at Wick. From Wiltshire and mid-Surrey northwards, there was snow-cover on between twelve and fifteen mornings, rising to twenty in north London, twenty-two at Ambleside, and twenty-three at Colchester and Liverpool; only, however, eight at Redcar and three at Hastings; and there was none at all at Portsmouth, Dorchester, Guernsey and Pembroke.

Was Christmas 1981 a 'white Christmas', or not? Those who betted on it found that it was not, since no measurable snow fell on the roof of the London Weather Centre on the 25th, though in fact the ground was snow covered in many parts from London northwards until the 27th or later.

January 1982

On 5 January 1982, cold air surged south across Britain, leaving the Orkney Isles and the Isle of Lewis covered with 10 to 15 cm (4 to 6 in) of snow. Much of the south of England and South Wales had forty-eight hours of snow on the 8th and 9th. In Surrey the depth was about 15 cm. Further west, snow depths increased to 30 to 40 cm (12 to 16 in) over the Cotswolds, South Wales and Herefordshire. At Newport, Gwent, there were 60 cm (24 in), at Carmarthen 70 cm (28 in) and at Swansea 40 cm (16 in) (Fig. 131). Many towns and villages were cut off for several days, travel being difficult, and, in some of the worst affected parts, travel was forbidden by the police. Deep drifts and abandoned vehicles looking like snow drifts had made movement too hazardous. The general snow-cover lasted until around 15 January, when a thaw without much rain melted the snow. In consequence, there was no serious flooding.

February 1986

Snow fell on twenty-four days at Reigate, Surrey, and on nineteen days in Nottingham. There was little disruption to transport because the few millimetres of powdery snow were easily blown off the roads by traffic. Gardens on north-facing slopes in Surrey were covered by 4 cm (2 in) of snow for most of the month.

January 1987

On the 10th a cold front introduced very cold Polar Continental air, accompanied by snow showers. More powdery snow fell next day. More was forecast for the following day, but it turned out to be very cold and sunny (page 172). On the 13th heavy snow fell from low cloud, thin enough for a pale moon to show through, at Coulsdon. There was an exceptionally heavy snowfall around the Thames Estuary. Leigh-on-Sea reported 44 cm (17 in) of level snow (Fig. 132). Southend and Gravesend were seriously affected and featured on national television news. A general thaw began on the 20th. An unusual feature of this snowy spell were the very low temperatures accompanying snow: −5°C (23°F), or lower. Who says 'It is too cold to snow'? There were few trains on the third-rail electric system.

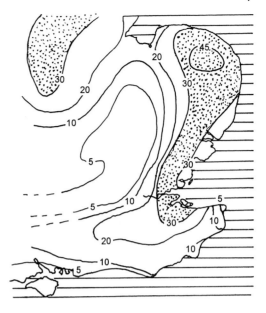

Fig. 132. **10–20 January 1987.**
Snow depths around the Thames Estuary and in East Anglia. Intense convection over a relatively warm sea and convergence of air were held responsible. As the cold air moved inland convection and snowfall declined rapidly. The difficulties of forecasting snow depths should be apparent from the 5 cm in Margate to 44 cm (17 in) in the Southend area.

8 December 1990, West Midlands

This was a superb example of wet snow: the opposite in snow quality to the 1987 snowfall. The affected area was about 50 kilometres (30 miles) wide and orientated north/south through the Midlands from west Oxfordshire to part of Humberside. Just after midnight around Birmingham heavy rain turned to heavy snow, accompanied by a strong wind. Within half an hour there was a complete covering. Had the precipitation been less intense the snow would not have settled significantly as the temperatures were fractionally above freezing. There is a saying 'Snow won't settle on wet roads.' It will if the fall is heavy, as this event shows. On the M6, to the northeast of Birmingham, many hundreds were stranded in their cars or sought refuge in the National Exhibition Centre. All surfaces exposed to the wind were covered. By evening level snow depths were estimated at 42 cm (17 in) at Acocks Green and 34 cm (13 in) at Edgbaston. Much damage was done to trees and shrubs as the snow clung to them in spite of the wind. Electricity was interrupted by heavy accumulations of ice on power lines. Some people were without power for a week. Transport was badly affected for three days. A slow thaw began as soon as the snow stopped falling. Within four days most of it had gone.

5–12 February 1991: 'the wrong kind of snow'

This now well-known phrase was used by a spokesman or public relations expert for British Rail to explain the delays to train services caused by the unusually powdery snow which found its way into the motors of newly designed electric trains, so that few commuters managed to reach London on the 8th. It was not the actual depth of snow which was responsible for the interruption to services, as was the case two months earlier in the Midlands. It is quite true that powdery snow is a rarity in lowland Britain; but the phrase suggests that nothing quite like it had ever happened before. Powdery snow fell widely in 1987 as just described; also quite widely in 1985, 1982 and 1979 to quote recent examples. The snow fell with temperatures as low as −4°C (25°F) so that the individual flakes were loose, small enough for powder snow. Even central London had a week of snow-cover. Intense high pressure brought an outbreak of Polar Continental air. Snow-cover ranged from fourteen mornings at Eskdalemuir and Durham, to thirteen at Nottingham, eleven at Hastings and ten at Lowestoft. St. Mawgan, Cornwall, had four mornings with snow-cover, Anglesey five and Dublin two, though the Outer Hebrides had no snow-cover, a normal event when snow comes from the east. Snow depths varied from 13 cm (5 in) in Guernsey to 30 cm (12 in) at Chelmsford and 51 cm (20 in) at Bingley, near Bradford. In the London parks there were 12 cm (5 in).

May 1993

Weather-minded people staying at Westgate-in-Weardale were hoping to visit the climatological station at Moor House (556 metres, 1825 feet) but a sudden fall of 30 cm of wet snow, on the 14th, made this impossible. In the uplands May snow is not so rare for 'the fall of the average temperature with height is more marked in May than at any other time of year'. At this height lambing time is about the first week in May, so the snow was most unwelcome. The weather station has associations with Professor Gordon Manley who did pioneering work in studying our climate.

Christmas Day 1993

There was a white Christmas in many places though not London. Uplyme, near Honiton in Devon, Bridgwater, much of the Mendips, the Cotswolds, Wolverhampton and Skelmersdale had enough snow to cover the ground: a centimetre or more. Had this much fallen on any other day it would not have merited inclusion in this book. Some four million people experienced a white Christmas. Quite a few people travel to relatives on Christmas morning. There has not been one 25 December morning this century when travel by road would have been very difficult or impossible in lowland Britain. The evening of 25 December and Boxing Day have produced problems. Some may cite Christmas Day 1970 in parts of Kent and East Sussex as being rather fraught! A friend of mine had to abandon his car and stay in Chiddingfold, Surrey, on Boxing Day 1962, for the snow was just too deep to drive a small car through.

6 January 1994

The forecast suggested sleet. Instead there was a fall of up to 10 cm of wet snow. The snow belt was about 50 kilometres (30 miles) wide and stretched from the Isle of Wight to the south Midlands, including the western suburbs of London. Traffic chaos ensued because of slippery roads and poor visibility and jams caused by accidents. The forecast failed purely because of the intensity of the falling snow. Had the precipitation been gentle then few snowflakes would have reached the ground, but because a large number did, they cooled the air through which they fell and made it easier for subsequent snowflakes to settle. The sky cleared soon after the snow stopped, allowing the snow to freeze, as otherwise it would have melted away before morning. It gave some places in the south their first January snow-cover for seven years.

NOTES TO CHAPTER 16

1. M.C. Jackson, 'Snow depths in the blizzards of 1881 and 1891', M.O. Library, Bracknell. L.C.W. Bonacina, 'Snowfall in the British Isles during the half century 1876–1925', B.R. 1927, pp. 260–87. P.C. Spink, 'Famous snowstorms 1878–1945', W. 2, pp. 50–4. G. Manley, 'Snowfalls in Britain over the past 300 years', W. 24, pp. 428–37.
2. C. Carter, *The Blizzard of '91*, David and Charles, Newton Abbot, 1971.
3. Revd H.H. Breton, *The Great Blizzard of Christmas, 1927*, Hoyten and Cole, Plymouth, l928. L.C.W. Bonacina, 'Snowfall in the British Isles during the decade l926–35', B.R. 1936, pp. 272–92.
4. Author's observations.
5. J. Hurley, *Snow and Storm on Exmoor*, The Exmoor Press, Dulverton, Somerset, 1977.
6. R.M. Poulter, 'April snow in southern England', W. 5, pp. 209–10. G.T. Meaden, 'Late spring snowfalls in southern England', W. 28, pp. 485–6.
7. L.C.W. Bonacina, 'Snowfall in Great Britain during the decade 1946–55', B.R., 1955, pp. 219–29.L.C.W. Bonacina, 'Chief events of snowfall in the British Isles during the decade 1956–65', W. 21, pp. 42–6. C.S. Lowndes, 'Snowfall, 1954–1969', M.M. 100.
8. M.W.R.
9. I.T. Lyall, 'The snowfall of 2 June 1975', J.M. 2, pp. 73–7. G. Manley, 'Snowfalls in June', W. 30, p. 308.
10. Observations by M. A. Town.
11. I.C. Hudson 'The Highland Blizzard of 27–29 January 1978', J.M. 4, pp. 37–47.
12. S.D. Burt et al., 'The blizzards of February 1978 in south-west Britain', J.M. 3, No.33, pp. 261–88.
13. G.A. Southern, 'A Winter's Tale', J.M. 14, pp. 409–14.
14. I.T. Lyall, 'The winter of 1978–79 in Britain, J.M. 4, pp. 304–7. I.C. Hudson, 'February (1979) in the Highlands', C.O.L. No. 107. D.F. Evans, 'February blizzards in the Peak District', C.O.L. No. 107. W.S. Pike, 'The heavy early-season snowfalls of 7–9 December 1990', J.M. 16, pp. 109–21. A.K. Woodcock, 'The blizzard that divided Birmingham', W. 47, pp. 139–40. J.F.P. Galvin, 'The long night shift at Birmingham', W. 46, pp. 231–33. C.R. Finch, 'Snow depths, 8 December 1980', W. 46, pp. 248–9.

17 | Glaze

At a warm front, the milder air rises above the colder air, forming clouds and rain. Sometimes it is possible for the air to be below freezing point on and near the ground, and above freezing at cloud level (Fig. 133). Rain then falls from the warm layer into sub-freezing temperatures, making surfaces icy as the drops freeze on contact (Figs. 134 and 135). Luckily, freezing rain (or glaze) is rare in Britain, although it is common in eastern Canada, occurring on average seven times a winter in Montreal.[1]

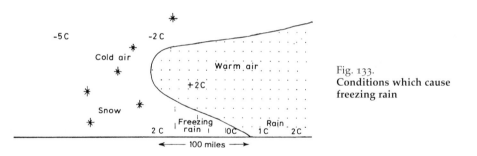

Fig. 133.
Conditions which cause freezing rain

Three things can happen to raindrops when falling from the mild air into the icy layer[2]: (i) they can remain liquid with a temperature above freezing, eventually turning to ice on the ground; or (ii) they can be supercooled, which means that their temperature is below 0°C, though they still remain liquid until they touch any surface, including wires and twigs, when freezing is virtually instantaneous and everything is encased in clear ice; or (iii) the raindrops can freeze on the journey down becoming clear ice-pellets, or a true 'frozen rain'. This is not the same thing as hail, described in Chapter 8.

Notable spells of freezing rain occurred in Britain in 1903, 1912 and 1927 in southeast England; in 1929 over Dartmoor; and in 1940, 1947, 1963 and 1966 in various parts of the south. Mid-Wales was affected in 1968 and parts of the Pennines and Sheffield in 1969. There was glaze over Wiltshire in December 1975, and over wide areas of England in February 1978. The south-east was affected in January 1979. One of the first documented instances of freezing rain in the London area was for 1903. It seems odd that the severe winters of the last century produced no reports of this phenomenon. Perhaps freezing rain made no news then since it caused less inconvenience and danger on the roads than it does today. There were also few wires to be brought down and no electric trains to be delayed.

Warm fronts moving into Britain from the south or south-west generally dislodge cold air in a matter of a few hours, once precipitation has begun. The most famous ice-storm of the century, that of 27 January 1940, was an exception. The rain was of the supercooled variety so that all objects exposed to the weather were glazed. Subsequently, snow fell on top of the glaze in some parts. Both remained on the ground until 4 February. The glazed area stretched from Kent to the Cotswolds and Exmoor, and from the South Downs over Cambridgeshire to the north Midlands. Damage was most severe in the area bounded by the Cotswolds, Exmoor, Salisbury Plain and the extreme west of Sussex. London only had snow with a few ice pellets. Fig. 134 shows telegraph wires on Salisbury Plain broken by the weight of ice. Further east, near Andover in Hampshire, the weight of ice on the wires amounted to 6 tons between posts, so it is not surprising that the wires were bent down to the ground. There were also many instances of wires and posts snapping under the strain.

'At Stonor Hill, near Petersfield in Hampshire, ice began to form about 4 p.m. on 27th January and continued the following day with a wind of between force four and five on the Beaufort Scale, so that all exposed surfaces were ice-covered and windows on eastern and southern sides of houses were stuck fast. During the night of the 28th–29th boughs of beech trees could be heard crashing down all night. The weight of the ice can be gauged from the fact that wires 1·5 mm thick had 30 mm (1·2 in) of ice on them. A similar story can be told of the Cotswolds. At Cirencester rain fell for forty-eight hours with the air temperature between −4° and −2°C (25° and 28°F). So complete was the covering of ice that all leaves on shrubs made a noise like castanets rattling in the wind, in an uncanny way. On the 28th it was almost impossible to walk on sloping surfaces because everything was so perfectly covered in ice.'[3] That night many places sounded like a battlefield. There was a constant roar of falling timber, just like a bombardment, which must have sounded menacing in that first winter of the Second World War. 'As a matter of fact the noise was out of all proportion to the damage, for the splintering of the ice-casing made even more noise than the rending of the wood; so that large limbs cased in ice made more noise than a whole tree normally does in falling, and especially at the moment of impact, when all the ice cracked and shivered like broken glass.'

A man at East Grinstead, West Sussex, seeing that it was raining, put up his umbrella only to find that no drips fell off; for they froze instantly. Many birds perished in the storm, and pheasants and rabbits could be caught by hand. There were numerous reports of birds and even cats being frozen to branches, and also of birds being brought down while in flight by the weight of ice on their wings. As can be

Fig. 134. **Telegraph wires on Salisbury Plain,** broken by the weight of ice from the freezing rain of 27–29 January 1940.

imagined, transport was badly dislocated. Driving a car was distinctly hazardous. A Captain Richards, stationed on Salisbury Plain, recounted that he could not get out of his car until help came, for the rain had frozen solid on his doors.

In the winter of 1947, there was widespread freezing rain in southern England, which fortunately did not last long. However, those who saw the trees encased in ice, lit by the early morning sunshine, will not forget the sight. In 1963, freezing rain fell on top of snow, making conditions very treacherous. Skiers and tobogganers on Box Hill near Dorking, Surrey, found the steep and curved slopes faster than usual and unpredictable.

On 20 January 1966 the evening weather-forecast warned people living in the south that a moderate to heavy snowfall was expected with strong winds and temperatures around or below freezing point. For a few days the wind had been easterly and quite strong, bringing day and night temperatures below freezing as the air was of Polar Continental origin. There was also a deep depression off the south-west approaches. The barometer had been falling steadily, hinting at a classic snow situation. The next morning many were surprised to see a damp world about them, which suggested that there had been a thaw and that there would be no trouble for travellers. Although freezing rain and drizzle had started before 8 h, most commuters managed to complete their journeys without much trouble.[4] But by mid-morning conditions were chaotic, and travelling by Southern Region electric train was third-degree torture because of the build-up of ice. Trains to Farnham and Portsmouth were badly affected in the evening rush-hour: a train would jerk forward, the lights going out and a brilliant flash from outside illuminating the darkened carriages; and it would then stop again with a jolt and the lights would come on momentarily. The process would repeat itself. One train actually welded itself to the conductor rail.

Accidents were double the normal on a winter Thursday. Quite a number were caused by people not realising the dangers (Fig. 135). There were, for instance, reports of people in Weybridge driving out of their garages in the usual way and failing to stop at the kerb as their cars skated gently on.

No reports came in of damage to power lines as it seems the rain was not of the supercooled variety, so that the raindrops would have fallen off the wires before they had had time to freeze. By the morning of the 21st, milder air had arrived and the cold spell was at an end.

On Christmas Eve 1968, a warm front was approaching Wales from the southwest. In front of it the snow turned to rain, then freezing rain.[5] The glaze was severe enough to break off 3 metres of Sitka spruce trees 11 metres high growing on the slopes of Plynlimon (Pumlumon Fawr). But this time the area affected was only sparsely populated. In March 1969, severe glaze occurred on the southern Pennines above 130 metres (427 feet) and over the Yorkshire and Lincolnshire Wolds.[6] This icestorm received widespread publicity because the Emley Moor television mast collapsed under the weight of ice on its stays. Unlike the 1940 storm, there were few reports of ice on the ground, although trees and wires were thickly coated.

Glazed frost developed over north-western Wiltshire and adjoining parts of Avon on 14 December 1975. At dawn, temperatures in the area were about −5°C (23°F). During the morning a warm front approached from the north and rain began falling from the warmer air aloft into cold air near the ground. By noon, between 3 and 5 mm of ice had built up on many surfaces.[7] Warmer air had melted the ice by the afternoon.

On 20 February 1978, fairly heavy freezing rain fell in Surrey for a time during the evening. All unsalted surfaces soon became coated with ice. By the next morning warmer air had arrived and there was little dislocation of traffic. In parts of the Midlands travelling was difficult on the morning of the 21st, and there were many accidents and delays; but by noon the most prolonged cold spell in the south since 1963 was at an end.

Fig. 135. **Glaze at Brighton.** Freezing rain or drizzle fell over much of south-east England on 20 January 1966. Road accidents were double the number usually reported on a winter Thursday. Many pedestrians were taken by surprise, too.

After a relatively mild day temperatures dropped sharply in the evening of 23 January 1979, and several hours of freezing rain occurred over a wide area of south and south-east England. As raindrops touched window panes or other objects, they made a distinctive pinging sound, freezing on impact to give about 3 or 4 mm (about 0·15 in) of glaze over all exposed surfaces on the ground. Later in the night the rain turned to snow. The next morning many people's routine journeys were made virtually impossible. Those who did venture out in their cars found the roads exceedingly slippery. There were numerous stories of people taking several hours to travel a few kilometres and inevitably there were many accidents.[8] (The situation was made worse by a strike of council workmen and on British Rail, though they would not have been able to operate anyway!)

Trees and telephone wires on the Blackdown Hills in Somerset were brought down by the weight of ice which built up between 10 and 13 February 1979.[9] Glaze was widespread over parts of southern England and the Midlands. Between 31 January and 4 February 1986 there was freezing rain or drizzle over the Pennines, parts of Yorkshire and the north Midlands. At Hemsby, the average ice thickness was 20 mm (0·8 in) above 300 metres (100 feet) with a bias on the windward side of objects. Burnley, Macclesfield and Buxton suffered interruptions to power supplies. Many farms were without electricity for several days. The TV news showed NORWEB crews chipping some 50 mm (2 in) of clear ice from downed lines near Buxton.[10]

Fig. 136. **Rime icing on Great Dun Fell in the Pennines**. Note the great thickness of ice on the mast.

On 27 December 1995 it seemed highly likely that snow would affect the south of England later in the week. As it turned out, an occlusion front moved slowly north on 30 December and freezing rain fell for about twelve hours, instead of snow, across much of the West Country, South Wales and the Midlands, as well as in Surrey, Sussex and Hampshire. At an occlusion front, as explained on page 29, the air in the warm sector of a depression is not in contact with the surface; so a temperature inversion will exist and rain may fall into the colder air and freeze. Roads and pavements became very slippery and there were many accidents, even though traffic was lighter than usual because of the danger. Hospitals had to cope with many fractures. A number of people said that they had not seen anything like it (that is, freezing rain) before. One person said that she had lived in Guildford for forty years and was somewhat taken aback when I mentioned the much more severe glaze in 1979, 1966 and 1963.

Between 1884 and 1903, meteorological observations were taken on the summit of Ben Nevis, at 1343 metres (4406 feet), at a specially built observatory which is now in ruins. Freezing rain was frequently reported during the winter, and also rime, which is a deposit of supercooled cloud droplets on a sub-zero surface: a phenomenon rarely observed in the lowlands. It had been rime as much as freezing rain that had been responsible for the collapse of the television tower on the Pennines in 1969.

NOTES TO CHAPTER 17

1. N.P. Powe, 'Freezing rain and glaze at Montreal', *Climatological Studies No. 15*, Ottawa.
2. D.W.S. Limbert, 'The formation of ice in mountains', *Mountaincraft*, London 1955, 27, pp. 14–18.
3. M.O., 'Glazed frost of January 1940', *Geophysical Mem. No 98*. C. J. P. Cave, 'The ice storm of 27–29 January 1940', Q.J. 60.
4. G. Parker and A.A. Harrison, 'Freezing drizzle in south-east England on 20 January 1966', M.M. 96, pp. 108–12.
5. J.K. Pedgley, 'Snow and glaze on Christmas Eve 1968', W. 24, pp. 480–5.
6. J.K. Page, 'Heavy glaze in Yorkshire, March 1969', W. 24, pp. 486–95.
7. 'Glazed frost in Wiltshire', J.M. 1, p. 136.
8. Author's observations.
9. A. Austin, 'The significance of altitude in glazed frost formation', C.O.L. No. 107, p. 12. D. Evans, ibid, No. 108.
10. W.S. Pike, 'The glaze accretions of 31 January–1 February and 3–4 February 1986', J.M. 12, pp. 3–17.

18 | Sunshine and Shade

The Campbell Stokes sunshine recorder is the standard instrument for recording bright sunshine. As Fig. 137 shows, it is a very simple, if expensive, piece of equipment; a crystal glass ball mounted on a pedestal focuses the sun's rays on to a special card, burning a hole or, more accurately, a long thin slit proportional to the duration of bright sunshine. Another instrument, the Jordan, permits the rays to pass through a cylinder on to photosensitive paper, producing a trace which can be measured. Various other electronic devices are also in use.

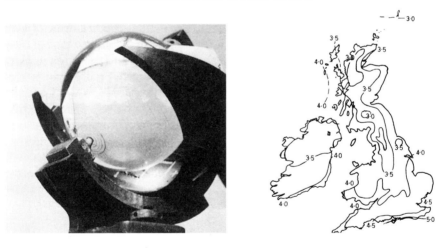

Fig. 137. *(left)* **The Campbell Stokes sunshine recorder at Kew Observatory**. The crystal glass sphere focuses the sun's rays on to card. The length of the burn mark on the card indicates the duration of bright sunshine to the nearest tenth of an hour. The burn mark can be seen just under the sphere.

Fig. 138. *(right)* **Average annual duration of bright sunshine in hours per day**

The theoretical maximum possible amount of sunshine in a year is 4380 hours. The average day from sunrise to sunset is twelve hours. If half an hour is deducted both after sunrise and before sunset, when the sunshine is not considered 'bright', the theoretical maximum is around 4000 hours. Totals approaching 90% of the possible occur in deserts like the Sahara: Britain's best, about 1800 to 1850 hours on average, occurs in the Isle of Wight and on the Sussex coast. Sunshine is especially welcome in Britain. Sunlight brightens up the mountains, enlivens the appearance of drab sub-urbs and makes even industrial landscapes look better. Photographers and snap-shotters also like to see the sun. By-laws prohibit building within a certain distance of existing premises, to prevent too much light being cut off. Houses with a southerly aspect can often command a premium over those facing north. More practically, the low angle of the winter sun can add substantially to winter warmth in a room. Farmland facing the sun usually commands a higher rent, because the growing season is longer than on the north side.

No doubt the reader can think of occasions when bright sunshine is unwelcome: it shows up dilapidated paintwork in the spring, and perhaps it spurs people on to spring-cleaning! It can also overheat some of our many-windowed office blocks and schools, as many office workers and schoolchildren could testify. Wet roads and a low-angled sun makes driving conditions dangerous. Sunbathing is certainly less fashionable than it was, because of fears about skin cancer; or is it that the sun-creams people use are not as good as they thought they were? The author, as a boy in the 1930s, was firmly told not to sunbathe for more than half an hour at a time.

The map (Fig. 138) shows the average number of hours of bright sunshine in a year.[1] The south coast is far and away the sunniest, being nearest to the continen-tal high pressure, and most sheltered from cloud-bearing westerly and easterly winds. North and north-west winds favour sunshine along the south coast. Sunshine totals decrease northwards and are lowest in the northern isles and over the mountains. Britain is not a sunny place compared to some parts of Europe. Nice has 2775 hours of bright sunshine a year; Barcelona 2487 and Naples 2396.[2] But then these places do not have a green countryside. Our yellow-brown grassy land-scapes of 1947, 1976 and 1995 very rapidly became green again. Nearer home, Brussels has virtually the same amount of sunshine as Kew. Paris is no sunnier than Worthing or Sandown.

In January, the south coast from Kent to Devon, the extreme tip of Cornwall and Pembrokeshire (Dyfed) are the most favoured regions, with a daily average of just over two hours. In June, the south coast from Dover to Weymouth and parts of west Wales have a daily average of over eight hours, or rather more than 50% of what is possible. The Outer Hebrides and Fife are the most favoured parts of Scotland. More details are given on the map (Fig. 139). A similar pattern prevails in July (Fig. 140), but the area with more than eight hours a day has diminished to include only the southern portion of the Isle of Wight. By August, the peak holiday month, there is still less sunshine. Once again (Fig. 141), the Isle of Wight is favoured, sharing the honours with Eastbourne. Seven hours of sunshine a day includes the time the sun is shining before we get out of bed in the morning. In June the sun rises at 4·42 am, in London, so the bulk of the population may well miss three hours of sunshine. In midwinter almost all of us will be up and about when the sun is above the horizon. Further information about sunshine is shown in Tables 33, 34 and 35.

Some Sunny Months [3]

June 1940 was a particularly sunny month (Fig. 142), sunshine being above average over all the British Isles. Totals exceeded ten hours a day along the south coast, in East Anglia and in North Wales and Lancashire. Only the northern tip of Scotland had less than seven hours a day, where moister winds off the Atlantic brought more cloud. In August 1947, places as far apart as Gorleston, Falmouth and Aldergrove, near Belfast, had over nine hours a day. This was another example of our normally green landscape taking on a yellow-brown look. July 1955 provides another example where a pattern of sunshine is different from the average (Fig. 143). Places around the Irish Sea shores had an average of over eleven hours a day, although the coast of Norfolk had only six hours, where winds blew off the cool North Sea. The summer of 1959 was the sunniest of the century at Kew. In June, Gorleston, Kew and Plymouth averaged over ten hours each day, though Stornoway was two hours a day below average.

In December 1962, pressure was high over the south-east of England, giving clear skies, which permitted the formation of radiation fog inland, while many coastal regions had beautiful weather: ideal for a walk along the promenade. Inland, hilltops such as Leith Hill or Box Hill were above the fog, affording a view across the 'murk' to the other 'hill islands'. This was the month of the fog and rime (Chapter 21), even though sunshine was way above average in the south-east outside the foggy areas.

Throughout February 1963, winds came almost uninterruptedly from the east or north-east, so that western coasts shielded by mountains were favoured. Sellafield, Cumberland, had a daily average of 4·8 hours; Valley, Anglesey, 4·7, and Ayr 4·5 (Fig. 144). Those who went to the Western Isles in August 1968 probably still have vivid memories of the long hours of sunshine there, particularly as they escaped dismal weather over the south of England (Fig. 145).

Among high sunshine totals in 1975 were daily averages in June of 11·65 hours at Jersey, 11·00 at Plymouth, 10·89 at Margate and 10·77 at Eastbourne. Kew, Cromer and Valley all had a daily average of ten hours. July was sunnier than average and August produced some high amounts, with numerous places recording over ten hours a day, among them 10·67 at Torquay and 10·21 at Brighton. The following summer, which will long be remembered for the drama of the drought, was very sunny in June, July and August (Figs. 146 and 147). In East Anglia skies were virtually cloudless between 23 June and 8 July, giving a daily average of over fourteen hours a day for the sixteen days. For the whole of June, Lowestoft had a daily average of 10·47 hours of sunshine and in July 9·97. By August, the winds were more easterly and the sunniest parts were in the south-west and along the south coast. Ilfracombe recorded 333 hours of sunshine (10·73 hours a day), which broke the record for the month. At Eastbourne, the total was an hour less. Other high averages were 10·66 hours a day at Portland Bill, 10·67 at Torquay and 10·55 at St. Helier, Jersey. By contrast, while the south-east was cloudless, the Outer Hebrides had one of the dullest Junes on record, being furthest away from the anticyclone and experiencing persistent west winds.

Fig. 139. (*left*) **Average daily sunshine (in hours) for June**
Fig. 140. (*centre*) **Average daily sunshine (in hours) for July**
Fig. 141. (*right*) **Average daily sunshine (in hours) for August**

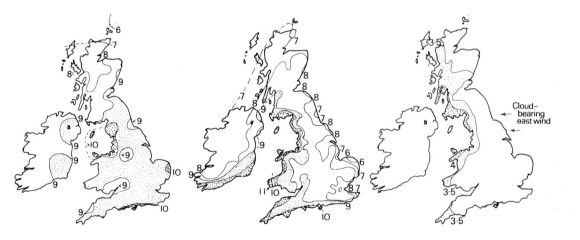

Fig. 142. (*left*) **Average daily sunshine (in hours) June 1940**
Fig. 143. (*centre*) **Average daily sunshine (in hours) July 1955**
Fig. 144. (*right*) **Average daily sunshine (in hours) February 1963**

Table 33 **Sunshine**

	Average monthly duration (hours)	Average percentage of possible	Average number of days with no sun
CAMBRIDGE			
January	52	20	12
February	69	25	8
March	117	32	6
April	153	37	3
May	192	40	3
June	202	41	2
July	186	37	2
August	177	39	2
September	138	36	2
October	105	32	5
November	57	21	11
December	42	17	14
Year	1490	34	70
FALMOUTH			
January	61	23	10
February	78	28	7
March	125	34	6
April	185	45	3
May	220	46	2
June	226	46	2
July	203	41	2
August	200	45	2
September	150	40	3
October	109	33	5
November	71	26	8
December	55	22	11
Year	1683	38	61
ABERDEEN			
January	56	24	
February	81	30	
March	109	30	
April	148	35	
May	182	36	
June	183	34	
July	159	30	
August	149	32	
September	130	31	
October	97	30	
November	60	24	
December	47	22	
Year	1401	30	

Table 34
Sunshine at Kew Observatory, London, 1881–1980; Royal Botanic Gardens, Kew, 1981–94[4]

	Average in hours per day (1941–70)	*Highest on record (hours)*	*Year*	*Lowest on record (hours)*	*Year*
January	1·54	92	1984	15	1885
February	2·28	115	1988	19	1947
March	3·62	183	1907	57	1888
April	5·41	245	1984	79	1920
May	6·56	320	1989	114	1932
June	7·14	302	1975	105	1909
July	6·34	334	1911	103	1888
August	5·90	279	1947, 1989	109	1912
September	4·77	221	1911	64	1945
October	3·29	160	1959	51	1894
November	1·92	103	1971	22	1897
December	1·38	72	1886	0·3	1890
Year	4·18	1931	1989	1129	1888

Table 35 **Extremes of Bright Sunshine**[5]

	HIGHEST ON RECORD			LOWEST ON RECORD		
	Hours	*Place*	*Year*	*Hours*	*Place*	*Year*
January	115	Bournemouth	1959	0·6	Cape Wrath	1983
February	167	Jersey	1891	4·3	Great Dun Fell	1966*
March	253	Aberystwyth	1929	25·0	Manchester	1916
April	302	Westbourne	1893	35·9	Manchester	1920
May	353	Worthing	1909	59·6	Great Dun Fell	1967*
June	382	Falmouth	1929	60·9	Crathes, Grampian	1912
July	384	Hastings	1911	49·6	Duartmore Bridge	1984
August	333	Ilfracombe	1976	43·9	Eskdalemuir	1912
September	281	Jersey	1959	34·3	Baltasound, Shetland	1967
October	207	Felixstowe	1920	8·0	Great Dun Fell	1968*
November	145	Falmouth	1923	3·3	Kinlochewe	1991
December	117	Eastbourne	1962	0·0	London	1890

* It seems likely that Great Dun Fell could well be the most sunless place where records are kept in every month of the year.

Fig. 145. **Average daily sunshine** (in hours) August 1968

Fig. 146. **Average daily sunshine** (in hours) June 1976

Fig. 147. **Average daily sunshine** (in hours) August 1976

Table 36 **The Best and Worst Summers at Falmouth and Kew this Century, using the 'Simple Summer Index of Bright Sunshine'**[6]

| **Best summers** | | | | **Worst summers** | | | |
| FALMOUTH | | KEW | | FALMOUTH | | KEW | |
Year	*Daily average sunshine May to September (hours)*	*Year*	*Daily average sunshine May to September (hours)*	*Year*	*Daily average sunshine May to September (hours)*	*Year*	*Daily average sunshine May to September (hours)*
1911	8·67	1989	8·40	1912	5·12	1912	4·76
1989	8·37	1959	7·92	1965	5·29	1931	4·82
1949	8·19	1911	7·88	1945	5·31	1968	4·82
1959	7·97	1976	7·74	1982	5·31	1981	4·87
1990	7·91	1990	7·37	1991	5·35	1916	4·91
1940	7·88	1933	7·29	1907	5·37	1932	5·00
1909	7·64	1940	7·21	1931	5·47	1946	5·06
1901	7·42	1975	7·16	1973	5·53	1913	5·10
1921	7·40	1929	7·15	1946	5·55	1920	5·12
1955	7·38	1949	7·09	1938	5·57	1954	5·16
1929	7·36	1928	6·97	1932	5·64	1910	5·18
1975	7·30	1934	6·91				
1934	7·30	1906	6·88				
1984	7·17						

In May 1977, daily average totals exceeded nine hours from west Wales to Ayrshire. Fair Isle, between Orkney and Shetland, had an average of seven and a half hours. The period 9–19 May 1980 was cloudless in many parts of the country. At Lerwick, there were eight days in succession when the sun shone for more than fourteen hours. On 18 May, there were 16·2 hours of sunshine; but there were eleven days in this month when there was only one hour of sunshine, or less. At Kew, the average for the eleven days was 12·7 hours. The remaining twenty days of the month had a daily average of just over four hours of sunshine. In spite of the dull beginning and end to the month some new records were broken. At Prestwick it was the sunniest May since records began in 1932. At Valley, Anglesey, the daily average for the month was just under ten hours.

Two useful ways of assessing a summer were described on page 155. Another and very simple method is the 'Simple Summer Weather Index',[6] which uses the sunshine totals for the five months from May to September. The reasoning behind it is well put by its originator, Trevor Baker, who writes: 'Whenever the sun is shining in summer it is acceptably warm; unless, early in the season, you are being attacked by a strong breeze from a cool sea and are stripped down to trunks or bikini.' In other words, sunshine figures give as good an indication of the standing of a summer as any other method (Table 36).

Both the Poulter and Optimum Summer Indexes take temperature, sunshine and rainfall over the three months, June, July and August; the Sunshine Index covers five months. The relatively poor positions of both 1975 and 1976, based on sunshine measurements, were due largely to the very dull Septembers in both years. It is interesting to note that 1976 did not rank among the ten best of the century at Falmouth and that 1959 was better at Kew, largely because of the long extension of summer into September.

In 1989 the four months May to August had an average of over nine hours' sunshine a day, beating 1959 and 1976 by a considerable margin in the London area. At Falmouth 1911 was better. Yet it is 1976 that stands out in the memory of the general public, because of the drought and heat. For poor summers the west has done rather badly in recent years with 1982 and 1991 fourth and fifth worst. At Kew 1981 was the fourth poorest: June, July and September averaged under five hours a day, and May a mere 3·6 hours, a very dismal performance.

NOTES TO CHAPTER 18

1. M.O., *Climatological Memorandum*, no. 72.
2. Met.O., 856c, Part III, Europe and the Azores.
3. M.W.R. (Figs. 142–7).
4. D.W.R. (Monthly Summary).
5. M.O., *Climatological Memorandum*, no. 72 (updated to December 1994).
6. T. Baker, 'Simple summer weather index', W. 24, pp. 277–80.

19 | Breezes and Gales

In Chapter 1 it was shown that differences in temperature, and hence in air pressure, will bring about a movement of air which we call wind. Over the British Isles, the prevailing wind blows from the south-west or west. The wind-rose (Fig. 148) shows the details for Kew Observatory.[1] It will be noticed that, although west and south-west winds are the commonest, the total of all the other winds exceeds the total for the so-called prevailing winds.

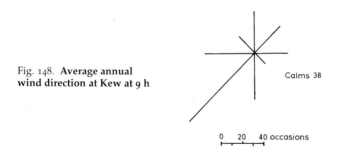

Fig. 148. **Average annual wind direction at Kew at 9 h**

Calms 38

0   20   40 occasions

In Cornwall, stunted trees, bent towards the east, show that west winds are strongest. But along the east coast winds from the North Sea are so frequent that the trees and shrubs at Skegness or Cromer are bent towards the west, suggesting that easterlies are more damaging than westerlies.

A systematic scale to measure wind speed or strength was devised by Admiral Sir Francis Beaufort at the beginning of the last century. It is still widely used today. People who listen to the BBC's shipping forecasts will be familiar with such phrases as 'west, backing south-west and increasing to 5', indicating a wind of force 4 increasing to force 5[2], with the wind direction shifting from west to south-west. For sailing enthusiasts, force 3 or 4 suggests good going. By force 5, the going is fast in the right craft, but things will happen too quickly, especially in crowded waters, for the inexperienced. If force 8 is mentioned, the haven of the clubhouse is better, or a walk along the beach to view the spray. The full scale is shown in Table 37.

Table 37 **The Beaufort Scale of Wind Force**[3]

Beaufort Number	Descriptive title	Effect on land features	Effect on insects, etc.	Range of speed in m.p.h.
0	Calm	Smoke rises vertically	Gossamer seen in air	Less than 1
1	Light air	Direction shown by smoke but not by wind vanes	Aphids fly; spiders take off	1–3
2	Light breeze	Wind felt on face; leaves rustle; vane moved	All species active	4–7
3	Gentle breeze	Leaves and twigs in constant motion	Hoppers, aphids, spiders grounded	8–12
4	Moderate breeze	Raises dust and paper; small branches moved	Beetles grounded; gnats, mosquitoes stop biting	13–18
5	Fresh breeze	Small trees begin to sway	Flies grounded, except horse and deer flies	19–24
6	Strong breeze	Large branches in motion, whistling in telephone wires	Moths and bees grounded	25–31
7	Moderate gale	Whole trees in motion	Butterflies and deerfly grounded	32–38
8	Fresh gale	Breaks twigs off trees	Only dragonflies still airborne	39–46
9	Strong gale	Slight structural damage to roofs, etc.	All insects grounded	47–54
10	Whole gale	Trees uprooted; considerable structural damage		55–63
11	Storm	Widespread damage		64–75
12	Hurricane			Above 75

For accurate measurements of wind speed, instrumental recordings are best. The simplest instrument is the cup-anemometer. Three or four cups are mounted symmetrically on a vertical axis, which rotates in the wind and is attached to a distance recorder. Another device has a tube containing a pressure-plate which is mounted on a wind vane so that it always points into the wind. The increase in pressure caused by the wind blowing into the mouth of the tube can then be measured. But for really comprehensive coverage of wind speed and direction an anemograph can be used. It must be mounted as far away as possible from the sheltering effects of buildings. A standard height above the ground is 10 metres (33 feet). The chart from an anemograph for the storm of 16 October 1987 is shown in Fig. 149.

Some facts about wind strength, using anemograph records over a ten-year period, are given in Table 38. In Chapter 3 it was said that the wind direction and strength at any particular place could be worked out with a fairly high degree of accuracy from a weather map with isobars on it; but, alas, this only holds true over the oceans or a bare plain. On land, the sheltering effects of hills, buildings and trees lower the wind speed for a distance of four to eight times their height, so that light winds may be blowing inland when there are gales on the coast. The high frequency of gales in coastal areas is notable: forty-eight days a year at Lerwick, thirty-three at Tiree, twenty-nine at Valley, twenty-eight at Scilly. By contrast Carlisle and Cranwell average two a year, and Heathrow only one. The windiest places, as the figures show, are in the north and north-west nearest to the tracks of Atlantic depressions and exposed to the full force of the wind off the sea.

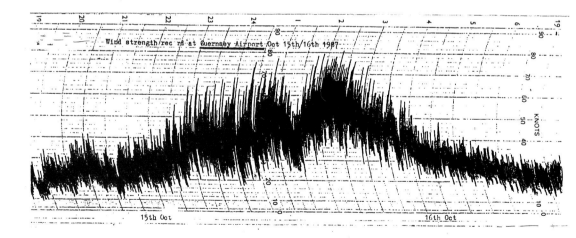

Fig. 149. **Anemogram for 16 October 1987.** The Great Storm or the so-called Hurricane.

At Stornoway, the wind blows at strong or gale force for over a fifth of the year, or a daily average of over four and a half hours. At Kew, there is strong wind for only five minutes a day over the year. Calm or near calm conditions are most probable well away from the sea or in sheltered valley sites. Thirty per cent of the year is virtually windless at Paisley and twenty-two per cent at Kew.

Strong winds and gales do damage to property, are unpleasant to walk or cycle against and rattle aerials and signs. Gentle breezes, on the other hand, are frequently beneficial, tempering the heat of summer, producing turbulence (or mixing of the lowest layers of air) so that fog does not form on a clear night, or helping to dry up the ground more rapidly after rain. Breezes lower the perceived temperature. Generally, the higher the temperature the more beneficial the wind. With temperatures near or below freezing, the intensity of the cold appears to be increased (as mentioned in Chapter 12).

A properly constructed and sited windbreak can make an appreciable difference to the local or micro-climate, since the wind speed can be substantially lowered. Apart from minimising wind damage to orchards and hop gardens, for example, a windbreak may also increase the air temperature in its lee by allowing the air to remain in contact with the ground for longer, so that plants grow more quickly. Less wind and slightly higher temperatures also help pollinating insects to do their useful work. Honey-bees will not fly in windy conditions. As Table 37 shows, Force 4 is the maximum wind bees are prepared to tolerate. Tiresome insects, such as mosquitoes and gnats stop biting if the wind is above Force 3. Those who like holidays in the west of Scotland should bear in mind that windiness has its compensations.

The windbreak can be considered as the first conscious attempt at weather modification. A loose windbreak, a hedge with gaps in it, is more efficient at slowing down damaging winds than a solid peep-proof one. A wall or thick hedge will often cause eddies (like those around some tall buildings, and described in Chapter 22) so that counter-eddies will be set up on the lee side of the windbreak, which may be as damaging as the full wind itself. Windbreaks strategically sited, sheep hurdles for example, can also prevent drifting snow covering roads in exposed situations.

Light breezes and moderate winds do not make news. Sometimes calms at sea may, when sailing boats are becalmed at a famous regatta. But gales 'hit the headlines'. An analysis of gales in each month of the year shows that January is the month with the greatest probability of days with gales, followed by December. Neither March nor September, the months of the equinoctial gales, are the most stormy. If a day with a gale is classified as having a mean wind speed of at least 39 m.p.h. (63 k.p.h.) for ten minutes, Lerwick has gales on 48 days on average; Tiree, 33; Holyhead, 29; Plymouth, 16; Jersey, 13; Turnhouse Airport at Edinburgh, 10; Aldergrove, 9; Elmdon, 3; and Heathrow, 2.

Table 38 **Percentage Distribution of Mean Hourly Values of Wind Speed During the Period 1950–59**[4]

	Force 8 & over >39 m.p.h. (>62 k.p.h.) %	Force 6 & 7 25 to 38 m.p.h. (40 to 61 k.p.h.) %	Force 4 & 5 13 to 24 m.p.h. (20 to 39 k.p.h.) %	Force 2 & 3 5 to 12 m.p.h. (8 to 19 k.p.h.) %	Force 0 & 1 <5 m.p.h. (<8 k.p.h.) %
Lerwick	1·3	14	44	34	7
Stornoway	2·5	17	43	24	14
Aberdeen (Dyce)	0·1	4	39	38	20
Bell Rock, Firth of Tay	2·7	22	47	20	8
Tiree	3·3	21	45	19	10
Paisley		1	22	47	30
Eskdalemuir	0·1	4	26	40	29
Point of Ayre	0·6	12	50	31	6
Carlisle		2	33	43	21
Durham	0·1	2	28	46	24
Holyhead, Anglesey	2·1	17	43	30	9
Southport	0·4	5	29	37	28
Ringway		2	32	48	19
Cranwell, Lincs		3	34	47	14
Birmingham (Edgbaston)		1	21	66	12
Cardiff (Rhoose)	0·1	5	40	46	9
Port Talbot	0·1	5	37	40	16
Heathrow		1	33	57	9
Kew		0·4	23	54	22
Shoeburyness, Essex	0·7	9	54	32	4
Thorney Island, Sussex		3	34	45	19
Plymouth	0·4	6	40	42	13
Scilly	1·2	14	46	28	11
Jersey	0·6	8	52	34	4
Aldergrove		3	40	45	13

The totals may not add up to 100% because of rounding and incomplete records.

Notes on Some Severe Gales[5]

28–29 October 1927

A deep depression off Ireland gave severe south-west gales over Ireland and north-west England. Very high seas along the Fylde coast of Lancashire caused flooding. Much damage was done to crops and trees by the large amount of salt spray carried inland. At Southport, the wind reached 153 k.p.h. (95 m.p.h.) in a gust and even the mean hourly speed was 113 k.p.h. (70 m.p.h.).

The Thames Flood of 7 January 1928

Serious flooding affected low-lying parts of The City, Southwark, Westminster and other riverside parts as far west as Putney and Hammersmith. Fourteen people were drowned in basements and a large amount of damage was done to property. The flood was the result of a tidal surge and a spring tide coinciding with high winds in the North Sea. On 6 January a depression near the Hebrides moved east-south-east to Denmark and strong winds, reaching gale force, blew from the north-west, forcing large quantities of water into the Thames Estuary. As a result of the flooding, loss of life and considerable damage to property, the walls of the embankment were made a metre or so higher. The Thames Barrier is expected to take the strain for many decades to come, though the south-east of England is slowly sinking; so causing an apparent rise of sea-level, most of which is due to the sinking of the land and not just to global warming and melting of the polar ice.

5–8 October 1929

Two intense depressions followed each other across Ireland. Gusts of 148 k.p.h. (92 m.p.h.) were recorded at Falmouth on four consecutive days. Scilly's gust of 177 k.p.h (110 m.p.h.) is the highest ever recorded in October at a lowland station.

Autumn 1935

Once again a depression moved across Ireland on 16–17 September, to give widespread gales. Even at Kew 111 k.p.h. (69 m.p.h.) was reached. About a month later, on 18–19 October, the northern part of the country had widespread gales. A Glasgow steamer was lost with thirty-seven lives. A mean hourly wind speed of 161 k.p.h. (100 m.p.h.) was recorded at Bell Rock, east of the mouth of the Firth of Tay.

Autumn 1938

Twice in 1938, on 4 October and 23 November, there were widespread gales. Kew achieved its record gust of 117 k.p.h. (73 m.p.h.), while at St. Ann's Head in west Wales the wind blew at an average speed of 129 k.p.h. (80 m.p.h.) for more than an hour.

April 1943

In 1943, north-west gales lashed Scotland and northern England, with a gust of 145 k.p.h. (90 m.p.h.) occurring at Manchester and Spurn Head. A number of longstanding records were broken, although the date of the storm was 7 April, not traditionally a windy month.

18–19 January 1945

There was extensive damage and loss of life in places as far apart as west Wales, Sussex and Suffolk. At South Shields the mean hourly wind speed was 105 k.p.h. (65 m.p.h.). The vigorous depression which brought the gales moved east-south-east across Scotland into the North Sea and was followed by the period of intense cold described in Chapter 14.

16 March 1947

After the severe winter a depression moving east-south-east brought gales to the whole country south of a line from the Mersey to the Humber. At Kew, 117 k.p.h. (73 m.p.h.) was recorded, a value reached on two other occasions. At Kingsway in central London (supposedly more sheltered) 121 k.p.h. (75 m.p.h.) was recorded.

1 February 1953: the storm of the East Coast Floods (Fig. 150)[6]

The gales previously described have been between the south-west and north-west, but this one came from the north. A deep depression moved east-north-east from south of Iceland and then turned rapidly south-east into the North Sea, while pressure rose rapidly to the west, producing a steep pressure gradient. In southern England the gales were accompanied by heavy driving rain and low cloud. In the North Channel, the Motor Vessel *Princess Victoria* sank with the loss of 132 lives. At Lerwick and Dyce, the mean hourly wind-speed was 121 and 105 k.p.h. (75 and 65 m.p.h.) respectively. This storm produced what is called a tidal surge or unusually high tide.

The tide tables published by the hydrographic department of the Admiralty state the time and expected height of the tides. The information given is correct and taken for granted by users. But sometimes an unusual combination of wind and air pressure will cause the tide to behave abnormally. In the centre of a depression, the sea surface rises because the atmospheric pressure is lower than normal. Theoretically the sea could rise about 30 cm (12 in) for every 33 millibars reduction in air pressure (or 1 in of mercury). Generally any increase in sea-level at one place is balanced by outflow of water. But if the wind is strong, the water shallow and the coastline constricting (as it is in the southern North Sea), then surges (or a build-up of water) may occur. At King's Lynn the predicted tide was 6·6 metres (21·6 feet) but the water came up 2·5 metres higher. The town centre was flooded by a sudden 2-metre wave. No one knew the flood water was coming as the noise it made was muffled by the roaring gale. At Sheerness, the peak tide was 2 metres higher than expected. Twenty people were trapped overnight in a coach between Sittingbourne and Sheerness and had to be rescued the next morning by an amphibious vehicle. There were power-cuts in Kent as floods affected a power station. On the Essex shore, most of Canvey Island's 13,000 people were taken to safety. The force of the wind and its unpleasantness can be judged from the photograph (Fig. 151). Over 300 people lost their lives in the four counties of Lincolnshire, Norfolk, Suffolk and Essex.

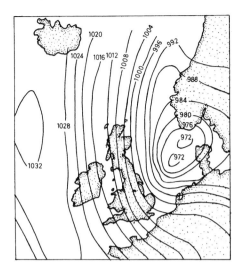

Fig. 150.
18 h, 31 January 1953. The East Coast Floods.
This was one of Britain's worst weather disasters.
Over 300 people lost their lives as a result of the
gales and floods, brought by a deep depression
moving south-east into the North Sea.

Fig. 151. **A scene in Sutton-on-Sea, Lincolnshire, on 3 February 1953,** when clearing up had begun.
The gale of 31 January deposited a metre or so of sand in the town. Eleven people were drowned and
the flood spread 3 kilometres (2 miles) inland.

26–30 November 1954

New records were broken in a prolonged spell of violent gales resulting from a series of intense depressions which moved northwards off Ireland. The South Goodwin Lightship broke her moorings and capsized; several other ships were wrecked. There were strong gusts inland and four small window panels of Salisbury Cathedral were blown in. High mean hourly wind-speeds and gusts included: 76 and 132 k.p.h. (47 and 82 m.p.h.) at Thorney Island in Chichester Harbour; 117 and 158 k.p.h. (73 and 98 m.p.h.) at the Lizard; 121 and 185 k.p.h (75 and 115 m.p.h.) at Brawdy in west Wales. That gust is a record for the whole of Britain for November.

29–30 July 1956[7]

This gale brought an abrupt end to a spell of fine warm weather. A depression deepened very rapidly as it approached the south-west of Britain. Mean hourly wind speeds of over 72 k.p.h. (45 m.p.h.) occurred widely, among them 105 k.p.h. (65 m.p.h.) at the Lizard, 95 k.p.h. (59 m.p.h.) at Shoeburyness and 89 k.p.h. (55 m.p.h.) at Aberporth on Cardigan Bay. The highest recorded gust was 150 k.p.h. (93 m.p.h.) at the Lizard. Even central London recorded a gust of 111 k.p.h. (69 m.p.h.). As would be expected there was widespread damage to sailing boats, crops, woodlands, and buildings. Summer gales are happily a rarity. But when they blow, the devastation wrought on plants and trees in full leaf is considerable because of their extra wind resistance. Many beautiful beech trees were felled by the gale in Arundel Park, Sussex; and about a hundred mature trees were blown over in the woods near the Goodwood racecourse. There were numerous reports of camps for schools, scouts and guides having to be abandoned after tents had been blown away. Many fruit growers suffered disastrous losses, and as much as three-quarters of the crop was wasted on some farms by nature's unwelcome summer pruning.

4 November 1957

An intensifying depression, which moved across Scotland, brought gales to the Midlands and south-east England. There was a gust of 148 k.p.h. (92 m.p.h.) at Dover accompanied by fierce rain squalls.

16 September 1961

A depression which began as a hurricane in the tropical Atlantic moved slowly along the western seaboard and caused one of the worst gales on record in Northern Ireland and north-west Scotland. There was a mean hourly wind speed of 93 k.p.h. (58 m.p.h.) at Tiree, with gusts of 167 and 171 k.p.h. (104 and 106 m.p.h.) at Ballykelly, near Londonderry.

11 January 1962

North-west gales developed in the rear of a depression which travelled east across Scotland and then turned north. The Midlands were severely affected. At Shoeburyness there was a gust of 132 k.p.h. (82 m.p.h.) and one of 167 k.p.h. (104 m.p.h.) at Hartland Point, north Devon.

15–16 February 1962 [8]

On 11 February a deep depression moved rapidly south-east towards southern Norway, bringing gales to Scotland and northern England. At Lowther Hill, Strathclyde (Lanarkshire), there was a gust of 198 k.p.h. (123 m.p.h.). After two rela-tively calm days, gales were renewed by an intense depression in the Norwegian Sea. A gust of 188 k.p.h. (117 m.p.h.) was recorded in the Shetlands and one of 190 k.p.h. (118 m.p.h.) at Lowther Hill. In Sheffield the maximum gust was 154 k.p.h. (96 m.p.h.). Over 7000 houses were damaged, three people were killed and about 250 injured. A number of schoolchildren found their houses severely damaged when they returned home from school. At Southwold, on the Suffolk coast, where a number of people lost their lives in 1953, the sea reached the high-water mark at low tide, which must have been a very disconcerting phenomenon. Luckily, the storm surge had passed when the time came for the normal high tide. So conditions were not as bad as they had been in 1953 because the surge and high tide did not coincide. The flooding and damage were worse on the other side of the North Sea. At Hamburg, the water rose to 4 metres (13 feet) above normal and over ⅓ metre (1 foot) above the previous record in 1825.

16 May 1962

At Benbecula in the Outer Hebrides, a gust of 161 k.p.h. (100 m.p.h) was recorded, which would be noteworthy in a winter month.

9 October 1964

An hourly wind speed of 117 k.p.h. (73 m.p.h.) was recorded in Jersey with gusts to 174 k.p.h. (108 m.p.h.). Much damage was done to glasshouses and the late tomato crop was destroyed.

15 January 1968

A deep depression north of Scotland moved to south Norway, bringing severe gales to the southern half of Scotland and north of England. There were mean hourly wind-speeds of 111 k.p.h. (69 m.p.h.) or more at Bell Rock, on Tiree and on the Forth

Road Bridge. On Great Dun Fell in the northern Pennines there was a gust of 214 k.p.h. (133 m.p.h.). The worst affected areas were in a narrow zone across the central Lowlands and southern edges of the Highlands, where many mature trees were snapped off. In the Glasgow area, twenty people were killed and thirty-eight injured. Over six hundred people were made homeless, fifty-three buildings collapsed and just under a thousand chimneys were damaged. High-tension cables across the Clyde were blown down, and shipping had to be halted until the cables were cut. In the North Sea the oil-rig *Sea Quest* snapped her anchor.

28–29 September 1969

Gusts over 129 k.p.h. (80 m.p.h.) were widespread over Scotland and a storm-surge occurred along parts of the east coast. A large part of Hull was under a metre of water. The chaos lasted nearly three hours until the tide turned and the flood-water ran back into the rivers. The floods cut off the old town's commercial area and shops and offices did not open that morning. Two electricity transformers caught fire due to short circuits. This surge made a tide over a metre higher than predicted. Later the same day the Humber ferries were cancelled because there was insufficient water in the channels.

2–3 January 1976[9] *(Fig. 152)*

After almost a week of stormy weather a depression moved across Scotland into the North Sea during the night of 2–3 January 1976, subsequently deepening and bringing very severe gales with winds of hurricane force or more, 119 k.p.h. (74 m.p.h.), during the evening between about 20 h and midnight. Among high gusts were 169 k.p.h (105 m.p.h.) at Wittering; 174 k.p.h. (108 m.p.h.) at Cromer; 164 k.p.h. (102 m.p.h.) at Norwich; 161 k.p.h. (100 m.p.h.) in Northern Ireland; and 122 k.p.h. (76 m.p.h.) on the roof of State House, Holborn, in London. The highest anywhere in the country was 216 k.p.h. (134 m.p.h.) at Lowther Hill, Strathclyde (Lanarkshire). Even the hourly average wind-speed was 158 k.p.h. (98 m.p.h.).

By 23 h every road out of Norwich was blocked by fallen trees. In the city itself over six hundred trees were blown down and many houses damaged. At Rugby, a large traffic sign was bent by the wind as if it were cardboard. The streets of many towns were littered with debris. Fortunately, not many people were out that evening to be struck by falling pieces of chimney-pot and roofing tiles. Structural damage was widespread in nearly every county, but it was most severe in a wide band from Northern Ireland across Lancashire and the Midlands to East Anglia. British Rail services were seriously interrupted near Rugby by fallen power-lines and trees on the track. At Liverpool a newly-built ferry was sunk, after being torn from its moorings. Altogether, damage over the country as a whole was estimated at as much as £100 million, a high enough figure to knock quite a sum off the shares of insurance companies on the Stock Exchange.

Generally the night sky was very clear, with only broken fast-moving clouds. But there were continual flashes, not of lightning as many supposed at the time, but caused by the thrashing and short-circuiting of high-tension cables by branches. Indoors, the electric lights flickered continuously even in some places in the south not so severely affected by the gales. It was also very difficult to try to ignore the screeching of the wind and put thoughts from one's mind of disasters to aerials and roofing tiles. Turning up the television set in many places was no help because that flickered, too. The power supply failed in many country places.

In Bushey Park, near Hampton Court, many fine large trees were blown down. Several growing not 30 metres apart were felled by the storm in such a way that their trunks were pointing in opposite directions, a sign of tornadic disturbance.

As with the storms of January 1953 and 1962, this depression produced a storm surge in the North Sea. By a stroke of luck, it did not coincide with the high tide, though the water remained above danger level for six hours. The depression also moved more rapidly than in 1953 so that there was less time for a surge to build up; and the winds were more westerly which helped. The improved sea defences were only breached in a few places, notably at Cleethorpes, where over a hundred homes had to be evacuated. In 1953, over 25,000 homes had to be evacuated. Sadly, four people died in Ireland as a result of the storm and twenty-four in Britain, mainly from falling trees and road accidents attributable to storm damage.

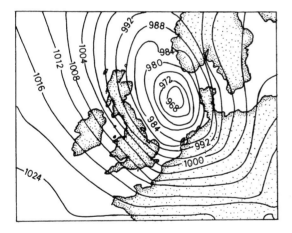

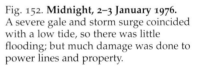

Fig. 152. **Midnight, 2–3 January 1976.** A severe gale and storm surge coincided with a low tide, so there was little flooding; but much damage was done to power lines and property.

10–11 January 1978

A rapidly deepening depression moved across England towards the Netherlands and severe northerly gales developed behind it. There were gusts of 153 k.p.h. (95 m.p.h.) at Tynemouth, 151 k.p.h. (94 m.p.h.) at Whitby and 130 k.p.h.(81 m.p.h.) at the London Weather Centre. There was widespread damage to property and woodlands, and numerous roads were blocked by fallen trees. Round the coast the piers at Margate,

Herne Bay, Hunstanton and Skegness were either badly damaged or destroyed. At Margate, the 150-year-old pier had been closed down in 1976 because it was thought unsafe, so the damage to all these piers is not necessarily a good indication of the severity of the storm (Fig. 153). The low pressure and strong northerly gale also caused a tidal surge (like in 1953), but this time the sea defences, which had been improved, held firm in most places. Only one person was reported drowned by flooding. It was estimated after the 1953 floods that the chances of a similar flood were about one in two hundred years. However, since then the water level has reached the same level in 1969, 1976 and 1978. The very slow downward tilting of the eastern half of England seems likely to make matters worse unless the sea defences are strengthened still further.[10]

13–14 August 1979: the Fastnet Storm[11]

On the night of 12–13 August a depression south-west of Ireland deepened rapidly and moved north-east. There were severe gales in many places, more especially in the South-West Approaches late on the 13th and during the 14th. As it happened, this was where the twenty-eighth Fastnet Race was taking place. Of the 303 yachts that left Cowes, 194 retired, twenty-three were abandoned and fifteen were lost because of the storm. Fifteen yachtsmen lost their lives. At noon on the 14th (Fig. 154) the wind was blowing at force 9 at Spurn Head on the east coast, and force 10 at Hartland Point, north Devon, though the worst period of the gales had passed and the skies had cleared (Fig. 155). Earlier, Mumbles, near Swansea, had recorded a mean hourly wind-speed of 93 k.p.h. (58 m.p.h.) with gusts to 111 k.p.h. (69 m.p.h.). Hartland Point had a gust of 137 k.p.h. (85 m.p.h.).

If the rough weather was the principal cause of the loss of life, there is also a suggestion that the more slender hulls of modern ocean-racing yachts are more easily 'knocked down' in stormy water. At the peak of the storm, many of the competitors were over the Labadie Bank. There the water shallows to between 60 and 90 metres, which was probably sufficient to make the waves even mightier and more confused.

This storm, which brought such tragedy, was a severe one for August, or for any time of the year, but it was not unprecedented. On average, gales can be expected on one day in the month at Valley, in Anglesey. At Scilly and in the Isle of Man they can be expected in seven Augusts out of ten. Off the south-east coast, Dungeness will expect a gale in one August in two. On 16 August 1970, Valley recorded a mean hourly wind-speed of over 80 k.p.h. (50 m.p.h.) and gusts of 100 k.p.h. (62 m.p.h.) were recorded at Plymouth and of over 129 k.p.h. (80 m.p.h.) at Kilnsea, Humberside. In the Isle of Man there was a gust of 130 k.p.h. (81 m.p.h.) on 25 August 1957, in association with a deep depression with a central pressure of 966 mb (28·53 in).

11 December 1981

The deep depression which brought blizzard conditions to much of southern England and a sudden thaw to places south of the centre of the depression, was also

Fig. 153. **The remains of Margate's 150-year-old pier after the gale of 11 January 1978**

responsible for flooding in the Bridgwater and Weston-super-Mare districts of Somerset. Gale-force south-east winds had held back the tide for some hours. A big turbulent wave could be seen offshore. Then a sudden veer of the wind to south-west, as the centre of the depression passed, allowed the waters to rush in, causing considerable damage.

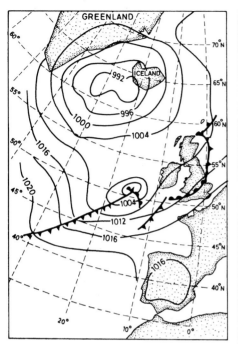

Fig. 154. *(above)* **Synoptic chart for noon,**
13 August 1979 during the Fastnet Race.
The depression south-west of Ireland deepened
rapidly as it moved north-east, and pressure at
the centre fell from 1000 mb to 984 mb (29·53 in
to 29·06 in) in less than twenty-four hours,
bringing winds of Force 10 to Hartland Point,
Devon, and Force 9 at Spurn Head.

Fig. 155. *(right)* **The cloud pattern at 15.30 h**
on 13 August 1979, about nine hours before
severe gales in the south-west approaches
wrought havoc in the Fastnet Race. The
depression west of Ireland shows up distinctly.

A resident of West Huntspill reported that he was seated with his wife in the lounge of his newly decorated farmhouse when the water first started to seep through the door shortly after 9.30 p.m. 'I managed to get through a 999 call and they told me the sea wall had gone. We rushed for the safety of the stairs, but inside no time at all the water was five feet deep.'[12]

16 October 1987: the Great Gale (the so-called Hurricane)

While watching the 12.55 BBC1 weather forecast on Sunday 11 October, I was impressed by the very large number of close isobars shown on the forecast chart for Thursday evening. I remarked to my wife that a phenomenal gale could be expected on Thursday night, if they were right. The Thursday evening was very wet and I had forgotten about Sunday's forecast. If Sunday's computer predictions had been kept, the forecast would have been more accurate. In many ways it was fortuitous that no warnings were given earlier in the evening as some people could have been caught by the worst of the storm, trying to rescue boats or shore up barns. Loss of life was minimal because the strongest winds came after midnight when few people were outdoors. A depression north-west of Portugal on the 14th deepened rapidly on the 15th off Spain and moved from Devon across to the Humber with central pressure below 960 mb (28.35 in). There were severe gales on the southern side of the track. Earlier it had been expected to take a more northerly track, but later a more southerly one affected France, which indeed suffered badly. There is photographic evidence at Versailles of many fine trees felled by the storm.

In the early hours many people were awakened by the noise of the howling wind and by bright flashes. These flashes were not lightning but were caused by short circuits. Power lines were thrashing against each other or were brought down by trees. The noise from my garden was frightening. Anything loose had become mobile: dustbins and their lids, flower pots, pieces of fence and loose tiles. I braved the elements to rescue a large potted chrysanthemum and was surprised by a rush of very warm air as I opened the door onto the garden, for the temperatures rose to 16°C (61°F) during the night, several degrees higher than in the early evening. Dawn revealed chaos in the south-east of England, roughly east of a line from Bournemouth to Cromer. The maximum recorded gust was 185 k.p.h. (115 m.p.h.) at Shoreham, Sussex. A stretch of the A246 near Leatherhead was blocked by over sixty large trees. People in country districts had no power for a week or more. Damage to property was considerable and local and national papers had pictures of badly damaged buildings and vehicles squashed by trees. Within the storm were tornadic disturbances, as there was evidence of trees being blown down in a swath through woodland, some lying to the left, others to the right. In mid-October most trees still had a full leaf cover, which increased the wind resistance. Another factor contributing to the damage was the waterlogged soil, so that trees were blown over more easily.

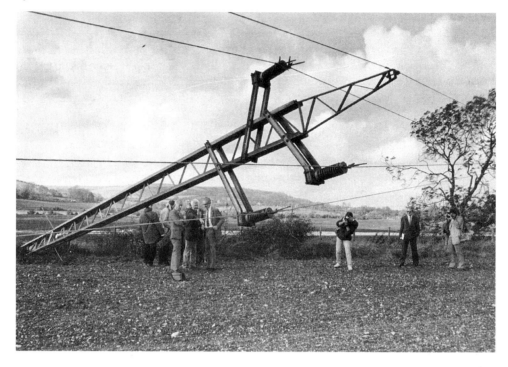

Fig. 156. **Scenes like this explain the widespread power failures as a result of the October 1987 storm.**
This is a 33000-volt line near Shoreham, Sussex.

It was undoubtedly a severe storm for any part of Britain, though northern Scotland suffers more severely more often. A 225 k.p.h. (140 m.p.h.) gust was recorded at Fraserburgh in February 1989, for example, but the publicity given to the so-called hurricane might have something to do with the fact that it happened on the doorstep of the media people in London, where much of the news originates on radio and TV and in newspapers, and of course, where more people can be affected. Ten million people being inconvenienced by a storm is much more newsworthy than a few thousand in a thinly populated part of the country. There is some evidence (see Chapter 22) that tower blocks may contribute to storm damage by increasing turbulence. Another factor in the perceived severity of the storm is our dependence on electricity, compared to fifty years ago. A storm such as this one would not have caused so much dislocation then. Few country dwellers would have had power to be interrupted, or freezers to be emptied. The fallen branches would also have been looked upon as a bonus by more people as a source of free fuel. Unusually the weather was responsible for the total interruption of commuter rail services into London from the south for the second time in a year—in January drifting snow had caused the trouble (see page 216).

25 January 1990: the Burns Day Storm

A deep depression with central pressure about 959 mb (28·32 in) crossed northern England and southern Scotland. Severe gales with gusts of over 161 k.p.h. (100 m.p.h.) were widespread to the south of the centre. Forty-seven people died in this storm. There were fatalities from Cornwall across to Clwyd and Cleveland—more than in the 1987 storm when the winds were stronger, though the affected area had been much smaller and the storm had occurred at night. Very severe gusts began soon after dawn in the south-west and moved eastwards, affecting Devon about lunchtime and the south-east soon after. At least thirty-five lorries and a double-decker bus were blown over. In Devon and Cornwall 101 caravans were badly damaged and thirty-nine destroyed. Because of its large surface-to-weight ratio a gust of 97 k.p.h. (60 m.p.h.) can easily overturn such a vehicle. The gale was severe in the south-east again but damage was less, partly because insecure trees had already suffered three years earlier.

   Both storms were a bonus for those supplying television aerials, garden fences and roofing tiles, though not for insurance companies who had to pay out on substantial claims. Severe gales were a general feature of the 1990 winter. Berry Head in Devon had a gust of 156 k.p.h. (97 m.p.h.) on 7 February. On 26 February there were widespread gusts over 145 k.p.h. (90 m.p.h.) in Scotland and over Ireland. In the Irish Sea the combination of low pressure, high tides and strong winds created a storm surge. There was much damage along the North Wales coast where the sea broke through the railway embankment and flooded many properties, with very little warning. Fig. 157 gives an indication of the devastation.

The 'Braer' Storm of January 1993

Pressure in the Atlantic at noon on 10 January was estimated to be 916 mb (about 27 in). This was the lowest recorded but there may well have been equal or lower readings in pre-satellite days. The oil-tanker *Braer* went aground on the southern tip of the Shetlands on 4 January. A succession of storms dispersed 85,000 tonnes of oil. Some of the oil was blown inland and polluted the pastures grazed by sheep. On Fair Isle, between Orkney and Shetland, there were twenty-two days in succession when the wind was gale force or higher. From 4 to 14 January the mailboat could not operate and many pupils were ten days late for school. Winds gusted to Force 10 on twenty-four days and to Force 12 on seventeen days. The winds blew from the west or south-west and, unusually, brought snow. The reason for this was that the depressions had formed at the margin of very cold air south of Greenland and much warmer air just to the south, the temperature gradient being much greater than usual.

NOTES TO CHAPTER 19

1. M.W.R., Table 4 (1963–72).
2. The word 'force' is not now used in forecasts. The omission of the word saves time.
3. Lyall Watson, *Heaven's Breath*, Hodder & Stoughton, 1984, p. 217.
4. M.W.R. (Annual Summary), Table IX, 1950–9.
5. R.O. Harris, 'Notable British gales of the last fifty years', W. 25, pp. 57–68.
6. C.K.M. Douglas, 'Gale of 31 January 1953', M.M. 82, pp. 97–100. K.F. Bowden, 'Storm surges in the North Sea', W. 8, pp. 82–4 and 100–13. H.A.P. Jensen, 'Tidal inundations past and present', W. 8, pp. 85–9. G. Reynolds, 'Storm-surge research', W. 8, pp. 101–8.
7. D.M. Houghton, 'The gale of 29 July 1956', M.M. 85, pp. 289–93.
8. C.J.M. Aanensen and J.S. Sawyer, 'The gale of 16 February 1962 in the West Riding of Yorkshire', *Nature*, 197, pp. 654–6. S.G. Irvine, 'High winds in Shetland' W. 21, p. 73.
9. C. Loader, 'Great gale of 2–3 January 1976', J.M. 1, pp. 273–83. M.S. Shaw and P.G.F. Caton, 'The gales of 2 January 1976', W. 31, pp. 172–83. G.T. Meaden, J.M. 1, p. 145. S.D. Burt, J.M. 1, p.146.
10. W. Ellsworth-Jones, 'Keeping the sea at bay', *Sunday Times*, 15 January 1978.
11. 'The Fastnet Race disaster, 14 August 1979', J.M. 4, pp. 277–80. 'The Fastnet Race inquiry', J.M. 5, pp. 15–22. *Observer* Colour Magazine, 25 November 1979.
12. *Bridgwater Mercury*, December 1981.

Fig. 157. **Floods in Towyn, North Wales, as a result of the storm surge of 26 February 1990**

20 | Some Local Winds

The Sea-Breeze

Everyone who visits the sea-shore will be familiar with the sea-breeze. A fairly common sequence of events along the south coast during a fine spell of weather is as follows: (i) the sea is an undisturbed mirror to the sky above until about half past nine in the morning. Then (ii) a gentle sea-breeze ruffles the surface of the water. The onset of it will be quite sudden, and (iii) it is quite likely the breeze will become force 3 or even 4 by the early afternoon, (iv) gradually dying away again by early evening. During the night there may be a gentle land-breeze (Fig. 158).

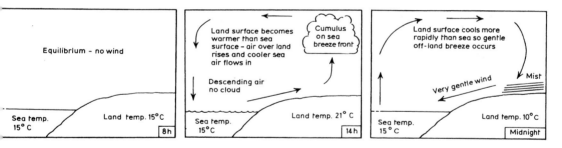

Fig. 158. **Sea-breezes and land-breezes**

A sunbather lying on a sloping beach looking up from sun-induced slumber will often see a line of cumulus clouds towering above the promenade or cliffs. Sometimes these clouds will grow large enough to look threatening (Fig. 7), but usually the clouds stay inland and only infrequently drift across to cut off the sunshine. As the afternoon progresses, the line of cloud may well appear to retreat as the sea-breeze advances inland. These clouds mark the line of the sea-breeze front (mentioned briefly in Chapter 1).

The Channel shore[1] is a good example of a sea-breeze coast, particularly from Dorset to East Sussex. Sea-breezes also occur in Devon and Cornwall, but the narrower peninsula and the high moors make the breezes less general. A sea-breeze coast can be described as a coastline where the difference in temperature between the land- and sea-surface is large enough to start the mechanism working. In fine weather, the coast will only be affected by light atmospheric winds off the shore. Using these criteria, the east coast from Northumberland to Essex is an even better example of a sea-breeze coast than the Channel coast, for the North Sea is much colder, thus increasing the temperature gradient, and (with the exception of the Yorkshire Moors) there are no hills to impede the inflow of cooler sea air. The parts of the North Sea and Channel coastlines regularly affected by sea-breezes are shown in Fig. 159. At Kilnsea,[2] on the Humberside coast, the dominance of winds from the east is striking (Fig. 160).

West-facing coastal places, such as Blackpool and Southport, are likely to be fanned by sea-breezes when the gradient wind is light from an offshore, or mainly easterly, direction. When the gradient wind is west, the breeze or wind will certainly blow from the direction of the sea, but it is not a true 'sea-breeze' for it is not a local wind, although probably to the uninitiated a sea-breeze will be thought to be a wind off the sea, whether it is local or associated with a large-scale weather system.

When pressure is low to the north of Britain and high to the south, westerly winds will prevail and will frequently affect the Lancashire coast by day and night. If the sky is fairly clear of cloud during the middle of the day, a temperature gradient will

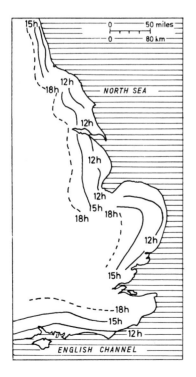

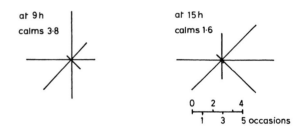

Fig. 160. (*above*) **The direction of the wind at Kilnsea, Humberside**, during August

Fig. 159. (*left*) **Penetration of the sea-breeze** along the east and south-east coasts. The figures show the average time the sea-breeze arrives.

occur and this will have the effect of strengthening the normal atmospheric wind during sea-breeze hours. The normal sea breeze working hours are 9.30 to 17.30 or 18 h, though both opening and closing times can be delayed or brought forward by an overall change in the weather pattern. Outside 'working hours', a wind off the sea cannot usually be brought about by a difference in the air temperature over land and sea but only by the general wind resulting from the distribution of high and low pressure.

Occasionally there is no wind at all by the sea. Such a situation is rare and comes about either when the strength of the land wind is balanced by the sea-breeze, so that no landward flow of cooler air can start; or when clouds cover the land and sea so that temperatures remain fairly constant. On 26 May 1974, lack of wind for most of the day frustrated all attempts to hold racing for competitors in an international regatta at Whitstable. 'Until three in the afternoon fleets of *Flying Dutchmen*, *Fireballs* and *Contenders* were drifting around in large interlocked groups; some boats point-ing this way, some that. Some had their sails up and others were under bare poles, but none of them moved except where the tide carried them.'[3] Sometimes Cowes Week, a highlight of the sailing year in the Isle of Wight, is similarly afflicted.

The Föhn (foehn)

This is a mountain wind: it often brings unseasonable warmth to the lowlands in the lee of the mountains. Some of the best examples of this wind occur in the Alps. 'Every winter comes the Föhn with its deep-toned roar, a sound which the Alpine dweller hears with fear and trembling, for there is danger of avalanches and fire in the tinder-dry conditions.'[4] Similar winds are experienced along the eastern flanks of the Rockies, along the north-east coast of Iceland and the coast of Greenland. The Chinook or 'snow-eater', as this wind is called, has produced phenomenal rises of temperature in a short time.

Britain's föhn winds are by comparison rather minor affairs for those who have experienced the real thing. A study of high winter temperatures in Britain shows that certain localities in the lee of mountains or high moors are sometimes favoured by abnormal warmth and low humidity, which is most noticeable during the winter. Aberdeen, the Moray coast, Cape Wrath, the North Wales coast around Llandudno, and less frequently Anglesey, as well as places along the Yorkshire coast to the lee of the moors - Whitby, for example - all have had high temperatures, 15°C (59°F) or more, in winter months.

What is a föhn?[5] To answer this question, it is necessary to consider the lapse-rate (or fall of temperature with height) of dry and saturated air. A moist Tropical Maritime air-mass would probably have a lapse-rate of about 0·6°C for each 100 metres (0·33°F for each 100 feet) of ascent. The Grampian Mountains in Scotland rise to about 1300 metres (4265 feet): on the summits, the temperature in a Tropical Maritime airstream would be about 2°C (36°F), with a surface temperature on the low ground to the west of the mountains of 10°C (50°F). As the wind descends the lee slopes, it warms up at the dry lapse-rate, for it has been robbed of its moisture. For every 100 metres of descent, it becomes not 0·6°C but 1·0°C warmer (0·55°F for every 100 feet). It arrives at the surface at around 15°C (59°F), very pleasantly mild for winter. Aberdeen has in fact recorded 15°C in January.

The explanation of winter warmth at Aberdeen is shown in the top diagram (Fig. 161). Föhn conditions also occur in dry airstreams, particularly those coming from the south-east. This type of föhn accounts for the high temperatures of 20°C (68°F) at Cape Wrath in March, and of 21°C (70°F) in November along the North Wales coast.

Sometimes föhn conditions occur at night. On 6 March 1977, the temperature at Red Wharf Bay, Anglesey, was 8·5°C (47°F) at 20.35 h. Mr K. Ledson, driving along the A5 in a car fitted with a thermometer to register the outside air temperature, found no change in the temperature until past Bethesda (Fig. 162).[6] At Tyn-y-maes there was a remarkably rapid rise and at Ty Gwyn farm the thermometer soared to 16°C (61°F) at 21 h. Mr Ledson stopped his car there: the sky was clear and there was a strong warm southerly wind descending the snow-covered Glyders. The warmth was all the more remarkable, for the snow-line was only some 350 metres (1150 feet) above the road. Further along the A5 the temperature showed a rapid fall near the sharp bend which separates the Nant Ffrancon from the Ogwen valley to 13°C (55°F) near Ogwen and to 10°C (50°F) within 200 metres (650 feet). 6 kilometres (4 miles) further along the temperature fell to 4°C (39°F) under a clear sky and in almost calm air. A south or south-easterly airstream always blows strongly and very gustily down the Nant Ffrancon from Ogwen to die away just before Bethesda. It does not always have föhn characteristics, but when it does, the maximum temperature is often just a couple of kilometres south of the Tyn-y-maes motel.

The Helm Wind

The northern Pennines in the neighbourhood of Cross Fell form one of England's largest stretches of high land over 760 metres (2500 feet). To the west, there is a steep drop to the Eden Valley which carries the M6 and the main-line railway to Scotland. Beyond the valley lie the mountains of the Lake District. The slopes and the sky above to the east of this valley frequently display an interesting phenomenon known as the 'helm wind'.[7]

It has been noted in many parts of the world that when a wind blows at right angles to a range of mountains, the wind speed increases substantially on the lee slope (providing the slope is reasonably smooth) and across part of the lowlands before dying away. Fig. 163 illustrates this. An ideal day for the phenomenon is a moderate, stable north-north-east to east wind over the Pennines. The wind-flow is intensified as it descends the steep 600 metres (nearly 2000 feet) escarpment to the Eden Valley near Penrith. It ceases abruptly, though its roarings still can be heard, only to rise again a few kilometres away as the normal rather gusty north-easter. The 'helm' can be recognised by the helm or 'helmet cloud' over Cross Fell and a similar roll of cloud called the helm bar or just 'bar' (Fig. 164), about 8 to 10 kilometres (5 to 6 miles) to the west.

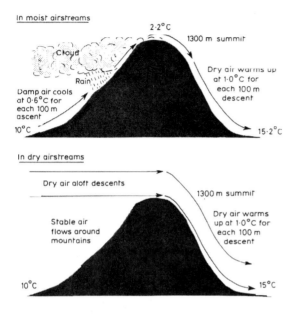

In moist airstreams

2·2°C

1300 m summit

Cloud

Dry air warms up at 1·0°C for each 100 m descent

Rain

Damp air cools at 0·6°C for each 100 m ascent

10°C

15·2°C

In dry airstreams

Dry air aloft descents

1300 m summit

Stable air flows around mountains

Dry air warms up at 1·0°C for each 100 m descent

10°C

15°C

Fig. 161. **Two possible explanations of föhn winds**

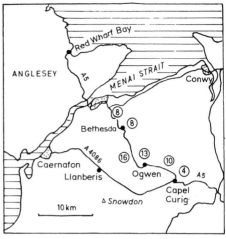

Red Wharf Bay

ANGLESEY

A5

MENAI STRAIT

Conwy

8

8

Bethesda

A 4086

16

13

10

Caernafon

Ogwen

4

A5

Llanberis

Capel Curig

10 km

△ Snowdon

Fig. 162. **Föhn conditions on the A5.** The figures in circles show the air temperature recorded at various points along the A5, between 20 and 22 h on 7 March 1977. The high value of 16°C (61°F) between Bethesda and Ogwen should be noted. It indicates föhn conditions.

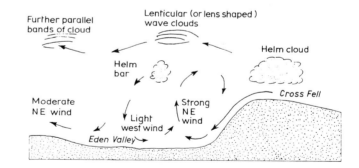

Fig. 163. **The Helm Wind of Cross Fell**

Lenticular (or lens shaped) wave clouds

Further parallel bands of cloud

Helm cloud

Helm bar

Cross Fell

Moderate NE wind

Light west wind

Strong NE wind

Eden Valley

Fig. 164. **Helm Bar**

NOTES TO CHAPTER 20

1. A. Watts, *Wind Pilot*, Nautical Publishing Co., 1975.
2. M.W.R., Table 4 (1971–5).
3. *The Times*, 27 May 1974.
4. G.E. King, 'Roar of the Foehn', *Conflict and Harmony*, George Philip, p. 31.
5. W.A.R. Brinkman, 'What is a foehn?' W. 26, pp. 230–9. I. T. Lyall, 'What is a foehn?' W. 26, p. 548. E.N. Lawrence, 'Foehn temperatures in Scotland', M.M. 82, pp. 74–9. W.D.S. McCaffery, 'Foehn effects over Scotland', M.M. 82, p. 151. J.G. Lockwood, 'Occurrence of foehn winds in the British Isles', M.M. 91, pp. 57–63.
6. K. Ledson, 'Foehn effects in North Wales', C.O.L. 83, March 1977, p. 12.
7. G. Manley, 'The Helm wind of Cross Fell', Q.J. 71, pp. 197–215.

Visibility—Sparkling Days
and Sombre Fogs

Seven counties can be seen from the top of Leith Hill in Surrey on a clear day, so it is said. The Isle of Man and Snowdon can be seen from Helvellyn; Ireland from Snowdon and Cader Idris; Skye from the summit of Ben Nevis; the Farne Islands from the Cheviots; and Dover from Calais. There is a snag though: the visibility must be right. All too often clouds sit on the summits, or haze blots out the distance. So lucky is the person who climbs to a viewpoint and has a good panorama. For not only must the visibility be good; the time of day must be right as well, because how far one can see depends on the position of the sun relative to the object. On the Pennines, the chance of a really good view is as little as one in twenty.

In the absence of any particles in suspension, visibility through the atmosphere would be about 240 kilometres (150 miles).[1] But there are always some particles present to scatter the light rays and cut down the view. Particles of dust and smoke and water droplets are so small that they only settle at extremely low wind-speeds. The dust and smoke produce haze, and the water drops mist or fog. Table 39 shows the average number of mornings over ten years with visibility between various limits for a number of weather stations.[2]

Really clear air is quite frequent at low levels in the north and north-west and in many coastal places. At Wick, there are 125 days in an average year when it is possible to see more than 40 kilometres (25 miles). Distance from towns and industry helps to keep the air pure; also Polar Maritime air which originates in a dust-free region is commoner in the north and north-west. Tropical Maritime air originates in a dust-free area, too, but the stability of the air concentrates the dust in the lower layers.

Over the country as a whole, visibility is best in polar airstreams, not only because of their clean source-areas but also because turbulence will carry away industrial smoke and dust and bring down clean air from aloft. In polar airstreams, Canary Wharf tower can often be seen from the Hog's Back, a distance of over 40 kilometres (25 miles), even in midwinter. In summer almost any airstream will produce some very clear days but easterly winds are frequently hazy, for they are usually stable in their lower layers, having been cooled by contact with the North Sea so that a temperature inversion is caused and the 'lid effect' prevents the dispersion of dust by convection.

The data in Table 39 refers to ten years, 1982–91. Earlier information for the years 1960–74 shows that fog was then commoner in our cities at 9 h.[3] At Heathrow there were six foggy mornings and at Elmdon seven. In low-lying rural parts the incidence of fog has not changed a great deal. Really clear days (with visibility over 40 kilometres) have increased from fifteen to forty at Heathrow and from one to twelve at Ringway. If we take days when it has been possible to see more than 20 kilometres (12 miles), the improvement in air clarity in our cities is striking. In the earlier period there were only 27 days at Ringway, now there are 103, bringing the hills of North Wales into view more often. At Elmdon the totals are 51 and 116 and for Heathrow 77 and 147. In the 1980s the use of coal as a primary fuel had declined and so had the output of smoke. At the beginning of the 1960–74 period coal was still quite widely used. There is some evidence to show that the overall clarity of the atmosphere is declining, away from cities, as the products of pollution in the cities are dispersed by the wind and turbulence. From the North Downs in Surrey a brownish stain above the distant horizon can often be seen in relatively still weather. The stain indicates where London is, even though horizontal visibility may be good.

Table 39 Visibility (Average Number of Mornings over 10 Years, 1982–91)

	Up to 200 m (220 yd)	>200 to 1000 m (220 to 1100 yd)	>1 to 4 km (1100 yd to 2½ miles)	>4 to 20 km (2½ to 12½ miles)	>20 to 40 km (12½ to 25 miles)	Over 40 km (25 miles)
Lerwick	2	10	23	119	134	74
Wick	1·7	5	15	85	134	125
Aberdeen	1	4	22	104	126	108
Tiree	0·2	2	14	128	113	101
Renfrew	2	5	21	171	104	55
Acklington, Northumberland	2·4	6	32	109	104	112
Waddington, Lincolnshire	8	9	43	152	114	38
Elmdon (Birmingham)	4	8	40	186	90	26
Heathrow	3	5	43	168	107	40
London Weather Centre	0	4	35	185	118	26
Ringway	2	5	42	213	92	12
Holyhead	0·8	4	19	126	124	100
Aberporth	1	8	19	186	99	53
Rhoose (Cardiff)	1	8	33	194	109	20
Plymouth	1	5	29	196	101	33
Aldergrove	2	6	22	144	128	64
Jersey	4	9	33	125	154	41

By 15 h in the afternoon visibility is often better than in the early morning, as turbulence clears away the mist and haze. The number of clear afternoons almost dou-
bles, to 73 at Heathrow and to 50 at Ringway. The air is clearer in the afternoon at
Lerwick by 4 and at Wick by 9 days, not a very significant difference, therefore, well
away from cities.

The information about fog refers to 9 h. In winter it is quite likely that the fog will
be at its thickest then; but there is often fog about before most of us are up, early on
a summer's morning. Coastal places, too, may experience fog when the sea-breeze
brings it in. June and July are the foggiest months at Lerwick, with fog on seven days
(though that does not mean it lasts all day); from April to July there will, on average,
be fog on two days in each month at Scilly. October can be very foggy: Exeter averages six days. Generally November and December are most fog prone. Sometimes
early evening can be foggy; but increasing wind may remove the fog by 9 h. Over the
period 1960–74 Elmdon reported fog on seven mornings but there were thirty-four
days in the year when fog occurred at some time during the day.

Ramblers, fell-walkers and climbers are vitally interested in visibility. They notice
the clarity of the air because they hope for a good view from a summit or ridge. But
the threat of fog also concerns them because it poses personal dangers. Visibility is
less than 40 metres (130 feet) on fifty-five mornings a year on average on Great Dun
Fell, at a height of 847 metres (2779 feet), in the Pennines, for example. Townspeople,
on the other hand, take comparatively little interest in the visibility range, unless it
is foggy and the train is late or the drive to work slower than usual.

Technically, fog is said to occur if visibility is less than 1000 metres (1100 yards),
though a motorist who can see 900 metres ahead of him need not reduce his speed for
safety reasons. Fog is caused by a suspension in the air of small droplets of between
one and ten microns. If the air is clean the relative humidity needs to be around 100%.
In other words, the air is saturated and roads and pavements will be damp. But
when the droplets contain nuclei of hydrocarbons or salt, as they often do in an
industrial area or by the sea, the relative humidity may be below 90% and surfaces
will remain dry. The visibility in a fog depends on the size and concentration of the
droplets: if they are small or numerous, the visibility will be poor; if large and sparse,
the visibility will not be as bad. In a mist, the drops are fewer and smaller and visibility will exceed 1000 metres. In everyday usage, people speak of mist patches when
visibility in the patches is restricted to a few metres, which affords an example of
confusion between 'official' and ordinary speech.

If a thin layer of air in contact with the ground is cooled to its dewpoint, dew or
hoar frost will be deposited. For fog to form, the cooling needs to go on long enough
for the depth of saturated air to be sufficient to make a cloud resting on the ground.
There are four kinds of fog: radiation fog, which fills valleys; sea fog; thaw fog; and
hill fog.[4] Sea fog and thaw fog are also known as advection fog because they are
caused by an advection of warm air which is subsequently cooled by the underlying
surface. This is in contrast to radiation fog which is caused by cooling of stagnant air.

Radiation fog is most likely to form when the sky is clear of cloud, the wind is very

Fig. 165. **Helm Crag, Lake District, Noon, 24 December 1972.** High pressure, light winds and clear skies are ideal conditions for radiation fog and frost to form. Fog can be seen filling the valley below, but increasing cloud and wind have stirred the fog sufficiently to reveal fields covered in hoar frost. The valley is cold and foggy, affording a sharp contrast to the sunshine of the fells. The clouds on the summit have only recently formed after a clear night and indicate the arrival very soon of moister air from an approaching warm front, as the high moves away east.

light (force 1 or 2), and the air already fairly damp (Fig. 165). An anticyclone is usually nearby or overhead, as it was in the London fogs of 1952 and 1962 (Figs. 166 and 167). The clear sky permits loss of heat from the ground by radiative cooling. Thus, the lowest layers of air are cooled to their dewpoint first. A very light wind helps to stir the cold air so that the cold is displaced upwards more readily, although a strong wind will prevent such rapid cooling by stirring a deep layer of air. Finally, moist air is needed so that comparatively little cooling is required to bring about condensation. Sometimes the formation of fog is accompanied by a rise in temperature. Early in the night the temperature falls through radiative cooling. Once a fog has formed, it acts as a blanket, just as a cloud does (which is what fog is) so that further cooling is retarded.

Radiation fogs like those described are mainly a phenomenon of valleys and the lowlands. This kind of fog is most frequent during the long nights of early winter when it often lingers through the next day, as it did on 30 December 1972 when there was little wind over the Vales of Evesham and Oxford. The fog was dense and traf-

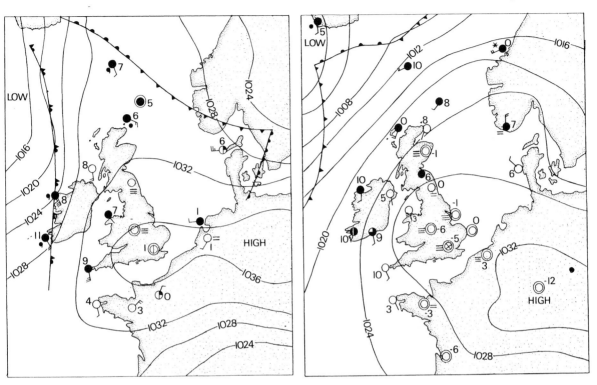

Fig. 166. (*left*) **18 h, 6 December 1952**. Over southern England pressure was uniform between 5 and 8 December 1952. In the London area, the fog was particularly bad, and over 4000 deaths were attributed to it. Some prime cattle were asphyxiated at the Smithfield Show and a performance at Sadler's Wells had to be abandoned because fog entered the auditorium.

Fig. 167. (*right*) **Midnight, 4–5 December 1962**. Between the 4th and the 6th temperatures remained well below freezing in the foggy areas and considerable rime deposits accumulated on trees and wires. Visibility was generally not quite as bad as in the fog ten years earlier, but the smell was worse.

fic was delayed, but those lucky enough to have ridge-top journeys had clear sunny weather. Some three-quarters of the way up the steep hill out of Broadway, Worcestershire, onto the crest of the Cotswolds, motorists found that an orange glow broke the gloom, like the bright fog-lights of approaching vehicles, as the sun began to penetrate the fog. Higher still, the sky was a clear blue and the fog could be seen as benign-looking white stuff in the hollows. These clear conditions were found between the top of Broadway Hill and Woodstock, a few kilometres north of Oxford, except at Moreton-in-Marsh, which as its name suggests lies in a hollow. Radiation fog is a menace on main roads and motorways when it is dense, because it is so often patchy, generally occupying the hollows, though sometimes hanging on the hillsides. It is possibly a little unfair to blame the fog for the multiple pile-ups which make headline news, for it may not be the fault of the fog. Drivers who disobey the

Highway Code by following too close to the tail-lights of the car in front and who do not anticipate the likelihood of fog are to blame. Fog is a hazard which every driver should watch out for whenever he is driving under starlight with little or no wind, particularly in undulating country in winter. The same applies to early morning day-light as well, as fog can form suddenly around sunrise. Even in summer, early morn-ing departures by eager holidaymakers anxious to get a good start before the traffic builds up, sometimes turn out to be not such a good idea after all when the fog fails to clear. The fast sections of the journey then become slower than anticipated, even at six on an August morning. But reference to Table 39 will show that the risk of fog is quite small. Sometimes 'the High Road' rather than 'the Low Road' can make the journey easier, although the reverse applies when hill fog is likely.

Fig. 168 shows the frequency of sea fogs along the east coast during August.[5] The view of the cliffs from Flamborough Head is a good one on a clear day. But many a visitor must have been disappointed, for fog is fairly frequent there compared with other places (Fig. 169).

Not uncommonly, the first hot spell of the year is spoilt by sea fog along the south coast if the wind is south-west, or along the east coast if the wind is east. Unlike radi-ation fogs on land, the sea fog is sometimes made worse by the breeze which can accompany it.

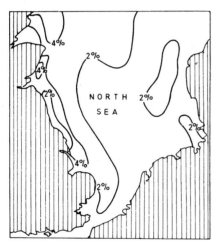

Fig. 168.
Percentage of sea fogs along the east coast in August

In August 1976 the fine weather was rather rudely interrupted in Edinburgh when sea fog (or haar) came in. Visibility at six in the evening was so poor that lights were necessary indoors. The view along Princes Street was reminiscent of the proverbial London fogs and the temperature was low enough for it to feel wintry, too. But the comparison ends there for the chatter of foreign voices indicated that it was the hol-iday season. Fortunately, the haar rarely lasts all day in Edinburgh. Lest any citizen of Auld Reekie should feel affronted let me add this, the same sort of sea fog could just as easily occur at Falmouth or Brighton, even if they do not call it 'haar'.

On land, advection or thaw fog is likely to form when mild air crosses cold ground, the cold surface chilling the already moist air to its dewpoint. The weather is dismal and raw, and in sharp contrast to the cold bright weather which it often replaces. As the name suggests this fog is often associated with thawing snow and raw cold.

Hill fog, as its name implies, affects the upper parts of hills and is really low cloud. It may occur at any time of the year, but is most frequent in winter when even low hills such as the Hog's Back, west of Guildford in Surrey, can be cloud-covered. Tropical Maritime air, which is usually moist, needs little lifting to be cooled to its dewpoint and it provides the most likely situation for hill fog to form. Sometimes hill fog makes everything very wet: minute drops hang in the wind and are caught by trees and other objects.

Higher hills and mountain tops like the Pennines are often foggy. Figures for Great Dun Fell for a five-year period, 1966–70, show fog on two-thirds of the mornings in a year; but in reality this is low cloud and unlike the valley fogs, for bad visibility can be accompanied by strong winds and rain or snow. By 15 h visibility has usually improved, so that just under half of the afternoons are foggy. No doubt similar, or even worse, visibility could be expected from mountain weather stations in some of the rainier parts, but there is no information available. On days when the moors are shrouded, roads over the high ground are best avoided by motorists if alternative valley routes are available.

Normally, when cold air sweeps over a warmer surface fog does not form unless the wind drops. Arctic sea smoke is common in the polar regions: the sea appears to be boiling, but the depth of steam is generally very shallow. It occasionally happens in British waters when the sea is 10°C (18°F) or more warmer than the air. A similar phenomenon can sometimes be observed over a wet road or roof, or on ponds, especially in the autumn when the first frosts of winter permit cold air to drain over their warm waters. Vapour from the surface of the pond is condensed but, because of turbulence, the saturated air rapidly mixes with drier air above and evaporates.

The word 'smog', meaning smoke fog, was first coined in the early part of this century, but was not widely used until the bad fogs of the 1950s. It is a feature of towns and industrial areas and is most likely to form when radiation fog also covers the countryside. When conditions are ripe for radiation fog a temperature inversion will exist so that a shallow layer of air near the ground will be colder than the air at, say, 150 metres (500 feet). And when this happens, smoke from chimneys rapidly loses buoyancy as it encounters the slightly warmer air. Owing to a great reduction in the use of coal and the passing of the Clean Air Act of 1956, smog is less common in many of our towns and cities than it once was. The decline could also have some connection with the decrease in rural fogginess resulting from more disturbed weather, and thus less propitious conditions for the formation of fog. It is likely, too, that the decrease in town fogs will influence the country nearby.

Information from Greenwich, Kew and Wisley shows a very marked decrease in fog at Greenwich and comparatively little change at Wisley (see Chapter 24). A fur-

Fig. 169. **Picture taken from a height of 1450 km (900 miles) on 7 July 1977 at 10.36 GMT by NOAA 5 satellite.** It shows an extensive bank of sea fog, stretching from the Tay to the Lincolnshire coast. Cold sea-water appears to be responsible for the fog, as the warm easterly wind was cooled to its dewpoint by the sea surface. Fig. 4 (page 5) shows the average sea-surface temperatures in August. The belt of cold water off the east coast should be noted. In fact, the fog almost exactly coincided with the cold sea.

ther set of figures for Stratford-on-Avon and Edgbaston indicates a similar if less obvious trend (Fig. 170), perhaps because Edgbaston lies on a plateau and London in a basin.[6]

Some Notable Foggy Spells since 1930

December 1935

Dense fog affected Scotland and parts of Ireland. Maxima were well below freezing in central Ireland on 20 and 21 December, reaching –2°C (29°F) and –3°C (26°F). In the Clyde valley there was a maximum of –9°C (16°F) at Paisley.

November 1936

Between the 19th and 28th of this month there was almost continuous thick fog at Manchester, with visibility under 200 metres for almost all the time, and as little as 5 metres on the 28th. In the streets the kerb was often the only object visible.

'After two or three days the fog and soot deposit began to get more and more definite and soon everything exposed to the fog was covered by a black wet slime, making even grass and shrubs appear drab. On one day a pilot flying over Liverpool in the fog, while trying to land at Speke airport, encountered slight icing, but the ice was almost black. '

The fog was also bad in other industrial and low-lying parts. An account from Bournville, part of Birmingham, says that the fog of 21–26 November 1936 was easily the most prolonged since observations began in 1911. At 9 o'clock in the morning of the 21st there was a bank of dense but beautiful white fog at ground level. So shallow was the fog belt at first that the sunshine recorder which is on a roof 24 metres (80 feet) high registered 3·8 hours of sunshine.

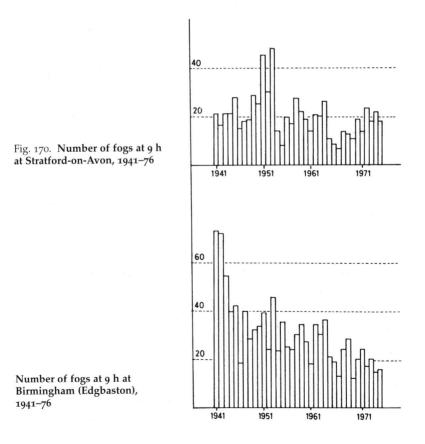

Fig. 170. **Number of fogs at 9 h at Stratford-on-Avon, 1941–76**

Number of fogs at 9 h at Birmingham (Edgbaston), 1941–76

'It was extremely depressing in the week which followed to see the gradual blackening of the fog owing to the accumulation of Birmingham's products of combustion. What slight breezes there were came from the north-east and aggravated the contamination of the air so far as Bournville was concerned. Rarely has there been in the Midlands a more striking need for smoke abatement.

'On the afternoon of the 26th, 1·8 mm (0·07 in) of rain fell. This had a hydrogen ion concentration of 3·0 which is remarkably acid for Bournville. During the week (beginning the 21st) copper and silver articles within doors acquired a lustrous black film within an hour or two of cleaning. Out of doors, metals, for example the copper on the roof of the rain gauge, displayed an unusual black iridescence. Pavements, roads, grass verges became sooty within a short time.' In the foggy areas, temperatures were often below freezing all day, though as high as 10° or 11°C (50° or 52°F) around the coasts.[7]

19–29 December 1944

There was widespread rime and fog in the south. In many places every object exposed to the air was covered with frost, which was spectacularly beautiful where the fog thinned in the afternoon and the sun shone. At Kew, Bristol, Hawarden near Chester, and Manchester the maxima were below freezing over Christmas. For most of the time pressure was high over southern Britain and low near Iceland, so that Scotland experienced milder Atlantic air.

5–9 December 1952 (Fig. 166)[8]

On the evening of 4 December the stratocumulus cloud-sheet broke up and allowed temperatures to fall sufficiently for fog to form. The next night saw a real water fog, the basis of all great London fogs, and, with temperatures falling below freezing, there was a copious deposit of rime even in the central parks. For the next three nights, visibility was so bad that movement even on foot in a familiar place was impossible. On the Isle of Dogs, near the Thames, it was reported that you could not see your own feet.

Bad though this fog was, it was not as unpleasant as the fogs that London used to experience, nor was it as bad as the fogs of 1936 in the Midlands and north-west described earlier. If our late twentieth-century fogs are cleaner, they may well contain far more dangerous lethal pockets. It is estimated that about 4000 people died as a result of the London fog of 1952. It was probably directly responsible for the introduction of the Clean Air Act of 1956, though even without this Act there would have been a substantial change from coal-burning fires to electricity, oil and gas, anyway.

January 1959

Fog was reported from some part of England on all but three days in this month. It was very persistent between the 23rd and the 29th in Lincolnshire where visibility was less than 25 metres for a considerable proportion of the time. Riding a motorcycle in London in the fog was dangerous and unpleasant: it was not possible to tell what colour the traffic lights were at a junction of the North Circular Road. Even with a makeshift mask the result of blowing one's nose left a black stain on the handkerchief for some days.

4–6 December 1962 (Fig. 167)

While visibility in London was not quite so bad as in the fog ten years earlier, there was a much more unpleasant smell which made the eyes smart. Over wide areas of the south of England there was a copious deposit of rime.[9] When a gentle breeze cleared the fog the rime fell off the trees and collected in little drifts.

The main market for coal is power-stations and very few factories and private households use coal, so the air in our cities is certainly cleaner in some respects, although pollution from motor vehicles is creating a more serious problem. The really thick city fogs are a thing of the past; but fogs still happen, even if visibilities are often of the order of 100 to 200 metres. Just before Christmas 1994 widespread fog produced much rime. Fig 171 shows the scene from Newlands Corner on the North Downs where warm sunshine had raised the temperature well above freezing but places in the fog zone were still below 0°C. Similar events occurred at Ringway on 31 December 1992 where the maximum was 0°C (32°F) but above the fog in the Peak District the temperature was a mild 10°C (50°F) with abundant sunshine. Numerous other examples could be quoted.

Fig. 171. **Radiation fog filling part of the Weald.** Looking south-east from Newlands Corner, Surrey, Christmas 1994.

NOTES TO CHAPTER 21

1. Meteorological Office, *A Course in Elementary Meteorology*, 2nd edition, HMSO, 1978, p. 58.
2. M.W.R., Table 4 (1982–91).
3. M.W.R., Table 4 (1960–74).
4. C.R. Burgess, W. 4, pp. 319–23. A.R. Meetham, 'Know your fog', W. 10, pp. 103–5.
5. F.E. Lamb, *Serial Atlas of the Marine Environment*, Folio 9, M.O.
6. M.W.R. (Annual Summary).
7. C.W.G. Daking, 'Unusual persistence of fog' (in Manchester and Liverpool), M.M. 71, pp. 252–3. F.H. Dight, ibid, pp. 253–4. S.K. Best, 'Dense fog at Bournville', M.M. 71, p. 235.
8. L.C.W. Bonacina 'Great London fog of December 1952', W. 8, p. 223. L.C.W. Bonacina 'An estimation of the great London fog 5–8 December 1952', W. 8, pp. 333–6. E.N. Lawrence, 'London's fogs of the past', W. 8, pp. 367–70.
9. Author's observations.

22 | City Weather

In the 1880s, the mayor of Middlesbrough opened the grandiose town hall of that thrusting new community with these words: 'Smoke is an indication of plenty of work (applause) - an indication of prosperous times (cheers) that all classes of people are employed (cheers). Therefore we are proud of our smoke (prolonged cheers).'[1]

Well, we are not proud of our smoke today. Indeed the Clean Air Acts have done much to increase the amount of bright sunshine in winter in our big cities and to lessen the frequency of fogs, as we have just seen. Contrary to popular belief, the centres of our major cities are not nowadays foggy places, and some desirable suburbs are far more likely to be foggy. This is partly because the rules of the Clean Air Acts do not apply there, but more importantly because the suburbs are less built up, making less stored heat available from the buildings and more moisture available from the ground, so that the air is more likely to be cooled to its dewpoint.

Much of this chapter is about the weather of London, simply because there is more information to hand, but observation suggests that what applies to the capital also applies elsewhere. Drivers leaving the West End in the late evening may well have no idea what is in store for them until at St. John's Wood or Wimbledon Common swirling fog spreads over the road, unless they have heard on their car radios that the motorway hazard warning lights have been switched on or, worse, that there has been a multiple pile-up. The city centre remains clear because the stored warmth from buildings and roads keeps the temperature up sufficiently for the air to remain above its dewpoint and no condensation occurs (though thick fogs can sometimes form when the air is only 90% saturated if there are enough nuclei, i.e. dirt, present). In the autumn and early winter, central London is often 'ringed with fog'. Later in the winter when the buildings have lost most of their stored warmth acquired in the summer, fog will spread further in, leaving a small clear central zone, which in a prolonged fog will also disappear, as happened in December 1962. Sometimes the fog is lifted up so that while horizontal visibility is good, the tops of tall buildings are shrouded and lights are needed indoors, and even in the street, by early afternoon.

Members of the jet set who expect to be deposited at Heathrow Airport but find themselves at Ringway, near Manchester, or Stansted, Essex, or even an airport on the wrong side of the Channel, may feel that cities are still foggy places in anticyclonic conditions. But the point is valid: they are not foggy in the old-fashioned sense, and Heathrow is a long way from the centre. Table 40 gives some details about the incidence of fogs. At Wisley, situated in open country 35 kilometres (20 miles) south-west of London, the number of fogs has not varied greatly in recent years and the small variations could be due to fluctuations in the number of ideal days for fog formation. At Kew, the fog frequency is down by well over half and the most dramatic change of all is at Greenwich, which was once a very foggy place but is now less likely to have fog than Wisley.

Table 40 **The Average Annual Frequency of Fogs around London Before and After the Clean Air Act**[2]

Years	Wisley	Kew	Greenwich
1934–43	30	34	41
1964–73	28	14	8

LA (Los Angeles), designed for the motor car, has been the textbook example for some decades of air pollution caused by strong sunlight interacting on the exhaust gases from motor vehicles. A strong temperature inversion, brought about by cooler sea air undercutting the very hot air over the land, provides an effective lid to rising exhaust gases. The pollution products rise and soon become trapped by the warmer air above, giving a thick haze, severe enough to cause eyes to smart and produce breathing problems. As car ownership spread, the newspapers and television showed pictures of traffic policemen with masks on in Tokyo, Mexico City and Athens, and even Paris in the great heat in 1994.

Since the manuscript for the first edition of this book was written in the late 1970s the number of motor vehicles in Britain has increased dramatically. So has the pollution in London and other cities and towns. There is some evidence that smoke pollution is increasing slightly with diesel engines being the source, but smoke pollution is nothing like it was in the days of steam engines on the railways and the widespread use of coal.

What weather conditions make for the worst pollution? High pressure is often the culprit. As the air descends from the edge of the troposphere it prevents the air near the ground from rising to a great height and dispersing. Some particularly bad occasions were just before Christmas in 1994 and in hot summer weather in 1995.

In the borough of Westminster the Department of the Environment's air-quality guidelines were exceeded for a total of thirty-nine hours in 1994.[3] Very simply, it is assumed that a similar amount of air pollution is emitted each day, it is the weather conditions that produce the high pollution levels. These conditions are hot, calm, sunny weather in the summer, and inversions (fog) in the winter. Wind can also bring pollution from outside areas; from neighbouring London boroughs or from Europe! High SO_2 concentrations are produced by easterly winds bringing pollution from the Thames corridor.

So though we are spared the dense smoke fogs we have the photochemical variety instead, which is probably becoming more common than the pea-soupers ever were. Just try walking over Waterloo Bridge in the rush hour. The traffic fumes leave a taste in your mouth!

Londoners sometimes look upon a visit to a friend or relation who lives in the country with apprehension. 'It's sure to be cold, wet and windy.' Such comments are not as wide of the mark as may be thought, for London is indeed warmer, drier and less windy than many other places. In fairly still conditions, the centre is often many degrees warmer than the suburbs, particularly at night.[4] Afternoon differences are not so great, although Weybridge had two successive days below freezing all day when the London Weather Centre reported maxima as high as 5°C (41°F) on 4 and 5 December 1976, for example.

This readily appreciable warmth in the centre has been called the 'heat island', and its presence is probable on any still cloud-free day or night, summer or winter. Only in windy weather does the heat island fail to show itself. On summer afternoons it might be expected that temperatures would be even higher in the centre than in the suburbs, or open country, but there is in fact comparatively little difference except on still days. Indeed in 1976 the highest temperature recorded, 36°C (97°F), came from Plumpton in East Sussex, and Cheltenham. Authoritative explanations as to why London does not always achieve the records for heat may well have something to do with the extra amount of dust or cloud, and turbulence brought about by tall buildings and possibly by an influx of cooler air from the Thames Estuary, carried by the sea-breeze. If daytime warmth is not spectacularly higher in heat waves, by night it certainly does show a different pattern with numerous readings over 20°C (68°F) occurring in hot spells, such as 21°C (70°F) at Earlsfield and 20°C at Greenford on 27 June 1976, and again in 1995 when Kew recorded two nights in succession over 20°C on 2 and 3 August.

We saw in Chapter 19 that gales or even strong winds are uncommon in London and in our other large towns and cities, though both Sheffield and Glasgow seem prone to severe gusts because of their location in the lee of high ground. It is also quite likely that city dwellers and office workers experience more days with gusty winds than the figures indicate, for there is evidence to suggest that very tall buildings create strong winds by funnelling the breeze down between the office block canyons (Fig. 172) like a man-made example of the helm wind described on page 256. Table 41 shows the number of hours when gusts exceeded 61 k.p.h. (38 m.p.h.) in five recent years at Heathrow, where the weather station is away from very tall buildings, and at the London Weather Centre in Holborn in the midst of tall blocks. Just one hour of strong winds at Heathrow in 1988 and nineteen at Holborn, a significant difference!

When gales are widespread, the differences are less, as in 1987 and 1990. There is the true story of a large block in Croydon which stood athwart the prevailing wind and created such severe counter-eddies that a shopping precinct was found impossibly uncomfortable by shopkeepers and customers, even when the gentlest of breezes were blowing elsewhere. Litter, leaves and dust were always on the move. The situa-

Table 41 **Number of Hours per Year with Gales at Heathrow and the London Weather Centre**[5]

Year	Heathrow	London Weather Centre
1991	1	12
1990	24	49
1989	9	28
1988	1	19
1987	12	23

tion was only remedied by the building of a roof over the precinct. Tall buildings also have other weather problems, created by the counter-eddies. Rain approaches from ground level, for instance, so that water seeps into walls from under window-ledges.

There is some evidence to suggest that large cities are prone to showers and thunderstorms, which are most likely to occur on the lee side on any particular day. The reason for this is the increase in turbulence brought about by the irregularity of the ground (woodlands and hedgerows give a much freer passage to the air than tall buildings) and also the extra warmth of city air. London seems to have its share of severe storms, such as the Hampstead deluge of 1975 (Fig. 48).[6]

Snow is not a frequent visitor to our major cities compared to the surrounding countryside. So much of our snow falls, as we saw in Chapter 15, when the air tem-

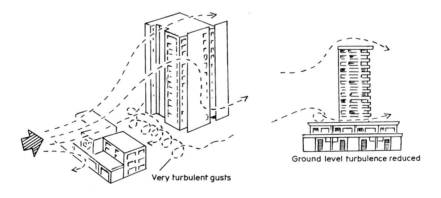

Very turbulent gusts

Ground level turbulence reduced

Fig. 172. **Airflow around tall buildings**

perature is around 0°C that the temperature of the ground is vital in determining whether the snow will settle initially until the surface has been chilled. Except after a prolonged cold spell, city roads and pavements will generally be too warm for snow to lie. On numerous occasions snow falls 30 kilometres (20 miles) out of London when none occurs at all in the centre. Sometimes there is simply no precipitation, sometimes just rain, not even sleet.

NOTES TO CHAPTER 22

1. B. Ward, *The Home of Man*, Pelican, 1976, pp. 36–7.
2. M.W.R. (Annual Summary).
3. City of Westminster, 'Air quality in Westminster', Annual Summary for 1994.
4. T. J. Chandler, 'The changing form of London's heat island', *Geography*, 46, pp. 295–307.
5. M.W.R. Table 7.
6. *The Times*, 'Climatology: London to blame for worst storms', *Science Report*, 17 February 1977.

23 | Our European Neighbours' Weather

Over 5800 paintings dated between 1400 and 1967 by major artists, whose works were in important art galleries, were once analysed to see what influence climate might have had on art. Table 42, based on this analysis, tells us quite a lot about the weather in various countries.

Table 42 **Relative Frequencies of Cloud Families by Schools of Artists** [1]

	High	Middle	Low	Convective
British	5	23	40	32
Flemish & Dutch	5	32	27	32
French	5	36	27	28
German	11	33	27	29
Italian	9	43	14	34
Spanish	6	38	28	28

In Britain, low clouds (stratus, stratocumulus and nimbostratus) are common, the stratus giving dull cheerless days and the nimbostratus rain. Convective or cumulus clouds are common in summer and in Polar Maritime air, as the artists have indicated. It is probable that the number of paintings with cumulus clouds exaggerates the total because these clouds would often be associated with good weather for artistic work. In the Low Countries, the cloud base is often higher, for the air tends to be drier than in Britain and this is reflected by the artist. Artists of the German School (which includes some Swiss, Austrian and Bohemian artists) were attracted by medium and high clouds, including wave clouds over mountains. The infrequency of low cloud in Italy is noticeable.

Another set of figures shows the frequency of cloudiness (Table 43). There was not a single British painting with a completely clear sky, and overcast skies were more frequent with British artists than with other schools.

As the table shows, paintings with clear sky in the background are in a minority in all of the schools. It is not surprising that no artist from the British school has painted a clear sky, for it has to be admitted that totally cloudless skies are not common in the British Isles; nor are they for that matter in the Low Countries and northern France, although there are a substantial number of sunny days, many coastal places having as many as seventy or more in a year when the sun shines for nine hours a day, or longer. Perhaps there is another factor. Many artists and their patrons like to have clouds in a landscape painting, especially to offset the gentle relief of Britain where cloud shadows give a sense of scale and depth. The desolate moors of Devon look better with some clouds and their shadows; and so do Salisbury Plain and the Cheviots, for example. At the other extreme, totally overcast skies are a feature of maritime north-west Europe, but are not common in Italy, and this is reflected in the statistics.

Table 43	Percentage Frequency of Cloud Families by Schools of Artists[1]			
	Clear	Scattered	Broken	Overcast
British	0	4	48	48
Flemish	7	11	44	38
French	9	10	49	32
German	16	16	42	26
Italian	12	24	42	22
Spanish	13	13	38	36

In Paris, the year's hottest day averages 34°C (93°F), a value reached only once every three years or so in Britain. Paris's absolute maximum of 40·4°C (105°F) has never been reached anywhere in the British Isles and is well above the unofficial record of 38·1°C (100·5°F) at Tonbridge on 22 July 1868. Paris's high absolute maximum is not particularly surprising since Tropical Continental air originating in North Africa does not get chilled there as it does by passage over the English Channel en route to us. Low temperatures are more extreme, even allowing for the urban site: nearly two degrees lower in Paris compared to Kew. Both places receive an almost identical amount of rain. Kew has 599 mm (23·6 in) and Paris 619 mm (24·4 in). The French capital has nearly three hundred hours of sun a year more on average than Kew, or the equivalent of thirty extra days with ten hours of sun—over a hundred hours more than Britain's sunniest spot, Eastbourne.

In Brussels, further away from the Atlantic influence and further north, both maxima and minima are lower. Rainfall is high in July and August. The annual total is a quarter more than at Kew. Sunshine is a mere twenty hours a year higher in Brussels (or the equivalent of two extra sunny days a year) compared to Kew; while sunless days total seventy-nine a year. In midwinter, afternoon maxima are noticeably lower.

Over a fifteen-year period (1961–75) the number of ice days at Strasbourg varied from nil to fifty-seven, with totals of twenty or more in nine of the years compared to an average of three in London (Table 27). In Paris the number of ice days is three times that of London; but Brest had ice days in only two winters.

An average over the years 1961–75 showed snow lying on four mornings at Brest, eleven in Paris and twenty-nine in Strasbourg, where the variation was from nil to sixty-three.[2] The highest total occurred in 1969, which was not a notably severe winter in Britain. In our severe winter of 1962–63, Paris had twenty-one mornings with snow-cover. Much higher averages and extremes occur in Germany, and other places to the east, where snow often comes with westerly winds of polar origin—especially in southern Germany where the altitude lowers the air temperature below the critical value. East winds of Siberian origin are very cold and dry and give both low maxima and low minima—as low as –10°C (14°F) at noon as far west as the North Sea coast of the Netherlands (Table 1). It is not surprising that standards of insulation and heating are better than in many English houses. They have to be!

Summer warmth increases rapidly away from the Channel and North Sea shores. Paris is three times as likely as London to have a warm day (25°C (77°F) or more); at Strasbourg, 'summer days' are four times as frequent. Fig. 173 shows the distribution of summer days.

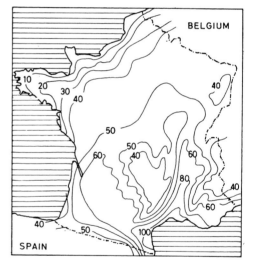

Fig. 173. **Average annual number of days with a temperature of 25°C (77°F) or more, over France**

Do our European neighbours have better weather than we do? It all depends on what is meant by 'better'. What is good weather for the tourist and hotelier can be frustrating for the factory or office worker. Bright sunshine does not please the angler, and many golfers prefer cooler summer weather.

If it is agreed that more sun is desirable, with hot days over 25°C (77°F) in summer[3] and cold, crisp frosty days with powdery snow in winter, then the continentals do better for weather. Once across the Channel, the weather should improve. But many a tourist will know only too well that it does not always do so. It can be cold and wet in Cologne in August, when north winds blow off the cool North Sea. In winter, the

ble 44 European Weather for the 30 Years 1961–90 at Selected Places. Temperature Data gives the Average (Mean) and the Warmest and Coldest Months in °C. *Rainfall Data (in italics) gives the Average (Mean) and the Totals in the Wettest and Driest Months in mm.*[3]

		Jan	Feb	Mar	Apr	May	Jun	Jul	Aug	Sep	Oct	Nov	Dec	Year
E BILT, NETHERLANDS, 2 m														
ean	°C	2·1	2·6	5·0	8·0	12·3	15·2	16·8	16·7	14·0	10·5	5·9	3·2	9·4
armest	°C	6·2	7·6	8·5	10·7	14·6	18·0	20·1	19·9	16·6	12·7	8·3	7·3	10·9
oldest	°C	−5·2	−3·6	2·0	6·1	10·0	13·4	14·7	15·2	11·6	7·0	2·6	−1·7	7·8
ARIS, FRANCE, 52m														
ean	°C	3·5	4·4	7·9	11·1	14·6	17·8	19·6	19·2	16·5	11·8	7·3	4·3	11·5
ean	*mm*	*54*	*43*	*32*	*38*	*52*	*50*	*55*	*62*	*51*	*49*	*50*	*49*	*585*
AMBURG, GERMANY, 16 m														
ean	°C	0·5	1·1	3·7	7·3	12·2	15·5	16·8	16·6	13·5	9·7	5·1	1·9	8·7
armest	°C	6·2	6·8	7·6	10·2	14·3	17·3	19·8	19·7	15·8	11·3	7·7	6·0	10·3
oldest	°C	−5·3	−4·8	−0·1	4·8	9·5	13·5	14·2	14·6	11·1	7·0	1·1	−3·5	7·4
ean	*mm*	*61*	*41*	*56*	*51*	*57*	*75*	*82*	*69*	*70*	*63*	*72*	*72*	*770*
ettest	*mm*	*134*	*86*	*149*	*115*	*122*	*185*	*169*	*167*	*153*	*147*	*137*	*173*	*991*
riest	*mm*	*8*	*8*	*19*	*8*	*8*	*16*	*5*	*8*	*11*	*16*	*19*	*12*	*541*
ERLIN, GERMANY, 50 m														
ean	°C	−0·2	0·8	4·2	8·6	13·9	17·4	18·8	18·4	14·6	10·0	4·9	1·4	9·4
armest	°C	5·4	6·6	8·3	11·5	16·3	19·7	22·1	20·9	17·7	12·2	8·0	5·6	10·9
oldest	°C	−7·2	−6·4	0·1	6·4	11·3	15·0	16·1	16·4	12·3	7·3	0·9	−5·5	8·2
ean	*mm*	*43*	*34*	*37*	*41*	*56*	*76*	*52*	*61*	*45*	*36*	*49*	*53*	*584*
ettest	*mm*	*119*	*83*	*117*	*88*	*124*	*186*	*121*	*160*	*110*	*144*	*89*	*130*	*784*
riest	*mm*	*12*	*4*	*10*	*1*	*4*	*11*	*7*	*10*	*5*	*4*	*11*	*10*	*385*
UNICH, GERMANY, 529 m														
ean	°C	−1·7	−0·4	3·4	7·5	12·1	15·4	17·5	16·8	13·6	8·6	3·1	−0·5	8·0
armest	°C	3·3	6·0	7·6	11·2	14·5	17·2	21·2	18·8	17·3	11·9	7·5	3·9	9·0
oldest	°C	−8·0	−7·4	−0·6	4·6	9·7	13·4	15·1	14·8	10·5	4·6	0·1	−5·7	7·0
ean	*mm*	*52*	*50*	*57*	*77*	*113*	*128*	*115*	*121*	*81*	*58*	*63*	*58*	*973*
ettest	*mm*	*125*	*137*	*141*	*153*	*255*	*255*	*217*	*243*	*168*	*196*	*159*	*133*	*1180*
riest	*mm*	*15*	*10*	*15*	*22*	*26*	*55*	*26*	*56*	*13*	*3*	*12*	*8*	*763*
ATWICK AIRPORT, UK, 62 m														
ean	°C	3·8	4·0	5·8	8·0	11·3	14·4	16·5	16·1	13·8	10·8	6·7	4·7	9·7
armest	°C	7·0	8·1	8·5	10·3	14·1	17·7	20·1	19·0	15·7	12·7	9·1	7·9	11·1
oldest	°C	−2·6	−1·0	2·5	6·3	9·8	11·7	14·7	14·5	11·5	7·5	4·3	1·5	8·3
ean	*mm*	*78*	*51*	*60*	*54*	*55*	*57*	*45*	*56*	*68*	*73*	*77*	*79*	*752*
ettest	*mm*	*183*	*143*	*118*	*106*	*142*	*123*	*87*	*131*	*224*	*196*	*196*	*146*	*1058*
riest	*mm*	*14*	*11*	*4*	*4*	*1*	*7*	*10*	*8*	*12*	*3*	*15*	*15*	*508*
NGWAY (MANCHESTER) AIRPORT, UK, 78 m														
ean	°C	4·0	4·1	5·8	8·0	11·3	14·2	15·8	15·7	13·5	10·7	6·5	4·7	9·5
armest	°C	6·8	7·3	8·7	10·5	13·5	16·9	19·4	18·5	15·1	13·1	8·5	8·1	10·5
oldest	°C	−1·4	−0·1	2·7	6·1	9·6	11·7	13·9	13·4	11·45	7·7	4·3	0·6	8·5
ean	*mm*	*73*	*54*	*61*	*53*	*61*	*69*	*66*	*83*	*78*	*81*	*83*	*82*	*843*
ettest	*mm*	*116*	*161*	*139*	*108*	*144*	*182*	*152*	*158*	*180*	*161*	*145*	*200*	*1073*
riest	*mm*	*15*	*4*	*14*	*4*	*11*	*16*	*13*	*7*	*11*	*26*	*40*	*15*	*673*

Continent has fogs, a fact which surprises some Americans who seem to think fog a purely English climatic affliction.

It is instructive to realise that Strasbourg can suffer really severe weather quite unlike anything we experience in our cities. In 1972, −23°C (−9·4°F) was recorded, and since 1961 there have been two other winters with extremes below −20°C (−4°F). All this as near to London as Dundee. Doubtless, French frost hollows, like their English counterparts, can produce sensational results.

The mean temperature figures in Table 44 show that at De Bilt in the Netherlands the average January is about 2°C colder than the south of England, the difference increasing in a severe winter to 3°C. Summer weather is fractionally warmer. Paris is marginally colder than the south of England in most winters and especially so when the cold comes from the east. Hamburg, near to maritime influences in Germany, has a January mean of 0·5°C (33°F) and in a severe winter is much colder than the south of England with −5·3°C (22·5°F) in January. July and April are driest in Gatwick but July and August have most rain in Hamburg, though convective rain will be common and so will not usually last long. At Ringway, November, August and December are the wettest months and April the driest. Southern Germany (Munich) averages three months with a mean below freezing, and as low as −8°C (18°F) in a cold winter. The months May to August have most rain; although there are no rain-fall-duration statistics available, it is a fairly safe assumption that the actual hours of rain are quite small and that the rain when it comes is heavy. Those who look out of a plane window while flying over southern Germany will often be rewarded with majestic cumulonimbus clouds with sharp anvils, indicative of thunderstorms and convective rain. Usually the flight will be high enough to avoid turbulence. Sometimes the air hostesses will say 'fasten your seat belts'. Winter is dry by contrast, with about half the summer totals, as high pressure so often keeps Atlantic depressions and their fronts away.

NOTES TO CHAPTER 23

1. H. Neuberger, 'Relative frequencies of cloud families by schools', W. 25, pp. 46–56. P.C. Spink, 'Climate in art', W. 25, p. 289.
2. Météorologie Nationale, Paris (1961–75 averages for France).
3. Met.O., 856c, Part III, Europe and the Azores, HMSO.

24 | Is Our Climate Changing?

Climate is often described as average weather, or what normally happens, including the run of early springs, bad summers, late autumns and occasional hard winters. Some people thought the climate had changed because we had two good summers in succession in 1975 and 1976. If this was true, it had changed again in 1977. Advocates of climatic change might point to the fact that July 1983 was the warmest month on record and the summer of 1989 was the sunniest in the London area. The early 1990s except for 1991, have been warmer than average. In other words, abnormally hot summers at infrequent intervals are a 'normal' part of our climate. In the same way, it is quite 'normal' when the south of England is gripped by an occasional severe spell, generally when the cold comes with an east wind, as when Reykjavik, the capital of Iceland, was found to be milder than London by several degrees in 1947, 1956, 1963.

Table 45 shows the thirty-year averages for monthly temperatures between 1901 and 1990 in six separate locations, and also the Central England Temperatures (often referred to as CET). These are composed of readings from more than one weather station, with allowances made for altitude and exposure. They form part of the CET record which goes back to 1659, and more reliably to 1715.

At Eastbourne, January means were 0·7°C (1·3°F) higher in 1901–30 compared to the following thirty-year period. The most recent averages show a half-degree rise for January, but the Eastbourne January temperature is still below that for the years at the beginning of the century. In the Highlands, the most recent thirty-year period with its January Braemar mean of 0·4°C (32·7°F) shows cooling in January. Summer months do, however, show increases in warmth of 0·6°C (1·1°F) at Sheffield in August and of 0·7°C (1·3°F) at Braemar. There is some evidence for global warming from the data in Table 45 if we look at the annual mean temperature. At Eastbourne the last thirty-year period is 0·3°C warmer than 1901–30 but there has been no change since then. At Kew the warmth has been fairly consistent: 10·1°C, 10·5°C and 10·6°C (50·2°F, 50·9°F and 51·1°F); but this is a composite record as Kew Observatory was closed in 1980 since it was considered to be less valuable as a long-period climatological station because of the influence of the expansion of the urban heat island. At Oxford and Sheffield and Braemar the warmest thirty years were 1931–60.

There was a run of milder winters in the 1920s and 1930s. In the next three decades the winters were much colder. Summer months show a reverse trend, suggesting an increase in continental air. In fact, the decrease of Atlantic influences has been very marked compared with earlier decades in the century, though we have no means of telling, at present at any rate, whether the trend is temporary or likely to be longer lasting.

One hears the older generation say, 'Winters were colder and summers more summery when I was young.' The records do indeed show that there was a run of severe winters by our present standards in the 1940s and 1950s, culminating in the notorious freeze-up of 1962–63. However, 1979 was a severe winter, especially in the Midlands and North. In 1981–82 there were very deep snows and record low temperatures. January 1982's mean was +2·3°C (36.1°F) over central England, which shows how statistics can hide information, as there were several nights in succession with night readings below –20°C (–4°F). The 1985 winter saw two very snowy spells and February 1986 produced continuous cold, with the month's warmest day reaching only 3°C (37°F) quite widely.

Four outstandingly warm sunny summers also occurred in the space of thirteen years, in 1947, 1949, 1955 and 1959. In the 1970s 1975 and 1976 were outstanding; in the 1980s 1983 and 1989. The 1990s began with record-breaking heat in August 1990; and 1993, 1994 and 1995 produced hot weather over much but not all of Britain.

In the 1930s there were two good summers in 1933 and 1934 but no severe winter, while the 1920s had the famous long, hot dry summer in 1921 and a very severe cold spell in 1929, which was the most prolonged since 1895.

It is certainly true that there was no snowfall of any consequence in many parts of the south from Christmas 1970 up to 1977, but records of snowiness only began about 1912, which is not long enough for us to be able to judge whether the climate is changing, or whether such a snow-free spell has happened before. The nearest approach since records began was in the 1920s when two winters were free of snowfall. Between 1987 and 1994 there was only one significant snowfall in the south of England: 'the wrong kind of snow' (see page 217).

Most people tend to remember the unusual and sensational events, and so it is with weather. Snow and heat are both sufficiently rare to make an impression. The author has heard it said that every day was fine and hot in June, July and August in 1976. A study of records shows that a few days were remarkably cool and cloudy, and even wet. The bad days were forgotten unless a special event was spoiled. Another frequently expressed view goes something like this, and shows some of the problems of relying on people rather than instruments: 'The winters always seemed snowier when I was a child. I remember it coming up to my knees.' But your knees were nearer the ground as a child! It might have been snowier for someone growing up in the 1940s, 1950s and early 1960s, but some winters had very little snow.

'We used to swim in the sea every day during the summer holidays, but recently the sea has been much too cold' implies that the summers of the past in the good old days must have been better. One cool August morning with the air temperature only

	Jan	Feb	Mar	Apr	May	Jun	Jul	Aug	Sep	Oct	Nov	Dec	Year
EASTBOURNE													
1901–30	5·4	5·2	6·4	8·3	11·8	14·1	16·2	16·4	14·6	11·6	7·7	6·2	10·3
1931–60	4·7	4·6	6·4	9·0	12·0	14·9	16·8	17·1	15·3	11·9	8·3	6·0	10·6
1951–80	4·9	4·8	6·3	8·5	11·8	14·7	16·5	16·8	15·1	12·2	8·3	6·3	10·5
1961–90	5·2	4·9	6·5	8·5	11·8	14·7	16·8	16·9	15·1	11·5	8·1	6·2	10·6
KEW OBSERVATORY, 1900–80, ROYAL BOTANIC GARDENS, 1981–90													
1901–30	4·7	4·8	6·2	8·4	12·5	15·0	17·1	16·5	14·1	10·5	6·4	5·2	10·1
1931–60	4·2	4·4	6·6	9·3	12·5	15·9	17·6	17·2	14·8	10·8	7·3	5·2	10·5
1961–90	4·6	4·8	6·7	8·9	12·4	15·6	17·6	17·2	14·7	11·5	7·2	5·4	10·6
OXFORD													
1901–30	4·4	4·6	6·1	8·2	12·1	14·4	16·6	16·0	13·7	10·3	6·2	4·8	9·8
1931–60	3·7	4·2	6·4	9·1	12·0	15·3	17·1	16·9	14·4	10·5	7·0	4·9	10·1
1951–80	3·8	4·1	6·1	8·6	11·9	15·0	16·8	16·5	14·3	10·9	6·8	4·9	10·0
1961–90	4·0	4·1	6·2	8·5	11·8	15·0	17·0	16·6	14·4	11·0	6·8	4·8	10·0
SHEFFIELD													
1901–30	4·3	4·2	5·4	7·5	11·1	13·6	15·7	15·3	13·1	9·8	6·2	4·6	9·2
1931–60	3·6	3·8	5·7	8·4	11·3	14·5	16·3	16·0	13·8	10·2	6·9	4·9	9·6
1951–80	3·6	3·6	5·4	8·0	11·2	14·3	15·9	15·7	13·6	10·5	6·5	4·9	9·4
1961–90	3·7	3·5	5·6	7·8	11·2	14·2	16·1	15·9	13·6	10·5	6·4	4·6	9·3
COCKLE PARK, NORTHUMBERLAND													
1901–30	3·2	3·2	4·3	6·1	9·1	11·8	13·9	13·6	11·4	8·5	5·1	3·6	7·8
1931–60	2·6	3·0	4·7	6·9	9·4	12·5	14·5	14·1	12·1	9·0	5·8	4·0	8·2
1951–80	2·7	2·8	4·4	6·5	9·3	12·5	13·9	13·9	12·1	9·4	5·5	3·8	8·0
1961–90	2·9	3·0	4·7	6·5	9·1	12·2	14·0	14·1	12·2	9·3	5·8	3·7	8·2
BRAEMAR													
1901–30	1·6	1·4	2·4	4·6	7·9	10·8	12·6	11·8	9·7	6·6	3·1	1·7	6·2
1931–60	0·6	1·0	2·8	5·2	8·3	11·4	13·1	12·6	10·2	7·0	3·9	2·0	6·5
1951–80	0·6	0·4	2·6	5·0	8·2	11·4	12·7	12·3	10·3	7·5	3·3	1·9	6·3
1961–90	0·4	0·5	2·5	4·8	8·0	11·3	12·9	12·5	10·2	7·3	3·2	1·7	6·3
CET (Central England Temperature)													
1901–30	4·2	4·6	5·6	7·6	11·3	13·8	15·8	15·2	13·1	9·8	5·8	4·5	9·3
1931–60	3·4	3·9	5·9	8·4	11·4	14·6	16·2	16·0	13·7	10·1	6·7	4·7	9·6
1961–90	3·8	3·8	5·7	7·9	11·2	14·2	16·1	15·8	13·6	10·6	6·6	4·7	9·5

14°C (57°F), a steady north-easter blowing and only fitful sunshine, a curious sight met the author's eyes on a north Devon beach, deserted save for two people, huddled in warm clothing, occupying deck-chairs. On approach, they felt an explanation was needed. Their two children were happily surfing in the sea, while their parents were unhappily 'freezing' on land. Would that summer day go down as a poor one weatherwise for those two children?

Fifty years ago, a higher proportion of the population lived in the country. They would have been more affected by ice and snow than townspeople. Now, the majority of us live in towns where heat from buildings and salt on the roads soon melts the snow, or prevents it from settling at all. Warmer houses and heated cars, trains and buses all insulate us from the cold. So it does seem as if winters have become milder and less of a hindrance. Those who drive long distances to work might not agree! The local authorities have achieved a good record in putting salt on the roads to melt any snow or ice, though I believe the sheer volume of traffic on the M25 on 5 December 1995 made it difficult to salt the roads.

A generation or so ago, many people spent the whole of their lives in one locality. Today, in this age of mobility, people do not stay in one place so much, and it is common for individuals and families to make several moves from one part of the country to another. To be brought up in Lincolnshire and to move to the Channel coast might lead one to make false comparisons when recalling past weather.

The average citizen who lives in a town (as Chapter 22 suggests) is less likely to see deep snow, or experience hard frost or fog, than his rural counterpart, although cities appear to be more prone to heavy rainstorms. Heavy rain falling over a concrete- and masonry-covered surface is more likely to produce serious flooding (Fig. 174) and cause inconvenience than a similar downpour in the countryside.[2] Also, more people will be affected and the storm will be more noticed. In fact, man has changed the weather and the consequences of weather locally, and he may be doing this on a world-wide scale, consciously or unconsciously.

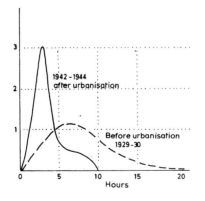

Fig. 174.
An increase in urban flood hazard.
The different response of the Silk Stream (which flows into the Brent Reservoir, off the North Circular Road) in north-west London to a rainfall of 10 mm (0·4 in) in one hour, before and after the catchment was urbanised.

Of the many substances entering the earth's atmosphere as a by-product of human activity, two, namely carbon dioxide (CO_2) and aerosol particles, including chemical products and dust, are of special importance, as possible causes of disturbances in the world climate, of which ours is a part. Molecules of CO_2 can remain in the atmosphere for several years, so that the gas becomes well mixed by the normal air circulation into all parts of the atmosphere.

A detailed report to the United States National Academy of Science in 1977 gave some very unpleasant predictions about the problems facing the inhabitants of the earth, if coal and oil were burned in ever-increasing quantities. Their main fear concerned coal. Coal reserves will probably last far longer than oil and the use of coal could have more impact. When it is burned in power stations, it adds to the mass of natural atmospheric carbon dioxide produced by volcanoes and other natural means. In recent years, the man-made increase has been at the rate of 0·7% a year. Until lately, the atmosphere was often treated as a vast self-cleansing dump for a whole variety of products, without apparent concern. At present rates of combustion of fossil fuels, the amount of CO_2 in the air will double in less than a century. It may well happen that by 2030 the world's average temperature will have increased by 2°C, bringing about a fairly rapid melting of the polar ice and causing alterations to the level of the sea and, increasingly, changes in the shape of the land, in more distant decades, as the earth's surface readjusts to the loss of the weight of the ice. The reason for concern is that CO_2 is almost transparent to sunlight but it absorbs heat. This creates a situation referred to as the 'greenhouse effect' so that the earth gets steadily hotter, like a garden greenhouse that admits the sun's heat and only lets it out slowly. The overall warming of the earth's climate between 1880 and 1940 has been ascribed to a 10% increase in the amount of CO_2 during this time. Since 1940, reversal of the trend has become apparent on a world scale, though this does not seem to have affected the British Isles markedly. The global situation suggests certain conclusions: either that the original proposition is wrong, or that there are other factors at work, such as variations in the incoming solar radiation or the slight wobbling of the earth in its orbit. It could also be argued that without the addition of man-made CO_2 the global climate might have warmed up less than it did up to the 1940s, and have cooled more rapidly since. It could even be that higher temperatures have led to more cloud and snow, so that more incoming solar radiation is reflected, resulting in lower temperatures.

It seems most probable that the result of an increase in the use of aerosols, apart from making blue sky less blue by scattering the light, will be a warming of the earth's surface by low-level particles such as dust and smoke, though high-altitude volcanic dust or clouds produced by planes flying at great heights are likely to have the opposite effect. In 1974, a BBC programme 'The Weather Machine and The Threat of Ice'[3] aroused great interest because it forecast a new Ice Age, with conditions deteriorating in the foreseeable future, so much so that world food production would suffer.

The Great Frost of 1683–84 was, historians tell us, the most severe that has ever visited these islands. Birds froze in mid-air and fell like stones to the ground. The mortality among sheep and cattle was enormous. But while the country people suffered the extremity of want and the trade of the country was at a standstill, London enjoyed a carnival of the utmost brilliancy. 'The Court was at Greenwich, and the King seized the opportunity to curry favour with his citizens. He directed that the river, which was frozen to a depth of twenty feet and more for six or seven miles on either side, should be swept, decorated and given all the semblance of a park or pleasure ground with arbours, mazes, alleys, drinking booths, etc. at his expense. Coloured balloons hovered motionless in the air. Here and there burnt vast bonfires of cedar and oakwood, lavishly salted, so that the flames were of green, orange and purple fire. But however fiercely they burnt, the heat was not enough to melt the ice, which, though of a singular transparency, was yet of the hardness of steel.' [4]

Frost Fairs also took place in 1795 and 1814.[5] There have been none since then, which suggests that the three worst winters in living memory, 1940, 1947 and 1963, were mild affairs by comparison. The CET monthly means for the three months beginning December 1813 were 2·8°C (37°F), −2·9°C (27°F) and 1·4°C (34·5°F). For comparison the means for December 1962 and January and February 1963 were 1·8°C, −2·1°C and −0·7°C (35°F, 28°F and 31°F) and yet there was no Frost Fair. It was not for lack of a sponsor, but for lack of ice on the Thames in central London. The effects of the London heat island cannot be held entirely responsible for this: direct warmth from the cooling water from power stations and a far more vigorous tidal flow, resulting from the demolition of the old narrow, arched bridges, and dredging of the river bed permitting an inflow of salt water are also responsible.

Between 1659 and 1994, there have been—if the accuracy of the earlier records can be relied on—thirty-seven calendar months with mean temperatures at or below freezing point. From February 1895 to January 1940, there was no month with a mean below 0°C. The 1870s had three very cold months, while the 1840s and 1860s had none, although the writings of Charles Dickens (1812–70) suggest that there was an abundance of hard weather in Victorian times. The last decade of the seventeenth century appears to have been the severest of all (using months below 0°C as a standard) when there were four such months.

The accuracy of the records depends on the quality and the correct exposure of the thermometer and the diligence of the observer in noting down the readings in the record book. The records quoted are thought to be of the highest standard. False readings can easily result from a standard thermometer screen if the exterior is not kept properly painted and clean. A dirty surface can easily absorb more radiation and thus affect the readings inside the screen. How important the colour of a surface can be to its heat retention properties can be simply demonstrated by placing a piece of white card and a piece of grey card of the same weight in sunlight and after a few minutes feeling the difference.

For some years now there have been temperature records from isolated places, and electronic recording is becoming more widespread as it saves labour. However there is a snag: the response time of digital instruments is much quicker than the mercury in a glass thermometer, so that maximum and minimum readings a few tenths of a degree higher and lower than those recorded by an ordinary thermometer may occur. Provided there is a long enough overlap between the traditional and modern there should be continuity of records. Taking a mean of 18°C (64°F) as a standard for a good summer month, there have been twenty-six such months from 1659 to 1994, with four of those in the last twenty years. Between 1677 and 1714, a lifetime in those days, there were no months over 18°C (64°F). Similar long spells without a warm month occurred from 1757 to 1783, and from 1868 to 1901. More recently, twenty years passed between the hot August of 1955, with a mean of 18·1°C (64·6°F), and August 1975 with a mean of 18·7°C (65·7°F). July 1976 had a mean of 18·4°C (65·1°F) and July 1983 of 19·2°C (66·6°F). Do these figures indicate changes of climate, or are they just part of the variety of weather patterns? Since the manuscript for the first edition of this book was written in the late 1970s the number of motor vehicles has nearly doubled in Britain and pollution has become a serious problem for society. Worldwide vehicle use has also increased enormously. The output of CO_2 has increased greatly in the last decade and we know that it traps some incoming solar radiation.

There are fears that the clearance of tropical rain forests may have world-wide climatic repercussions. Alteration of the colour of the earth's surface from a dark heat-absorbing colour to a lighter heat-reflecting one can lead to possible climatic consequences. The lighter the colour of a surface, the less heat is absorbed by the ground to give back more gradually to the air. In deserts, great daytime heat is often followed by relatively cool or even cold nights, as the shallow layer of very hot ground loses heat through the dry air very rapidly. Could what happens on the edge of the Sahara affect our weather?

There is also the rather intriguing science-fiction idea that widespread climatic change, initiated by drought,[6] could come about by oil spillages and the dumping of oil and plastic wastes in the sea, which might cover the surface of the oceans with a thin coating, reducing evaporation seriously, and causing fewer clouds to form to give rain. Surely the wind would break up such a plastic surface, though? In the disturbed weather of the temperate latitudes, this might well happen. But it seems equally likely that the anti-evaporation covering would remain intact in the equatorial belt of calms, and in the Horse Latitudes, where considerable evaporation now takes place. It may all sound far-fetched, but the quest for oil underneath the sea-bed has only just begun. There has already been some notable headline material for the press and TV, and plenty of film has been used to reveal the dramas of oil slicks.

It also seems reasonable to assume that if Britain was in the grip of an Ice Age 10,000 years ago, and if the climate is at its optimum now, the average annual temperature must have increased from around 0°C to about 10°C now. That suggests a warming of 1°C per millennium. But just as a trend towards increasing warmth, say in April, would not mean that each successive day is necessarily warmer than the preceding one, so perhaps it is with a millennium or a century.

The answer to the question 'Is our climate changing?' must be in the affirmative, though the changes will probably take longer than a human lifetime. But what would in fact be considered a change of climate? If the definition of climate given at the beginning of this chapter is the correct one, then any change of climate must be preceded by a sustained change of weather, significant enough to be noticeable in records over a long period. At the same time, the limit of upland cultivation would shift, climbing higher up the hillside if the overall weather improved, and coming lower down if the cold increased and the winters became more severe. Would our climate be thought to have altered if we were to have one severe winter like 1962–63 per decade? Has it changed because we have had several hot summers within a few years: 1989, 1990, 1993, 1994 and 1995? Don't forget, if you live in Cornwall or the west of Scotland, things may seem different.

It was said earlier that man has changed the weather in small but very significant ways in towns. But this is inadvertent climatic modification.[7] What of deliberate attempts to change the weather?

Numerous attempts at 'artificial rain-making', or causing it to rain in places where it would not otherwise do so, have been made in the United States and Australia. Carbon dioxide (dry ice) has been injected into clouds with the right characteristics, such as cumulus. The results have been inconclusive. But even at this stage there have been lawsuits because rainfall has, or has not, occurred where it should have done. It does not augur well for future weather control and climate modification.

Successful weather control could mean that the human race, or a number of its more knowledgeable members, would have a remarkable tool at its disposal. Accurate long-range forecasts could be issued to farmers, who would plant crops suitable for the forthcoming season. Unlike in the hot dry summer of 1976, when potatoes fared badly, and when they did swell as the result of the autumn rains, they could not be harvested very easily because the ground became waterlogged. Or there could be weather control so that conditions were always right for the production of basic food crops, at the expense of other desirable but less essential things. Would a standardised weather suit, for example, geraniums, hollyhocks and strawberries? The reader may well say that such ideas are nonsense. But man has altered the weather without trying. What could he do if he really tried?

One could foresee international friction, even war, between a tropical nation and a temperate zone nation, if, for example, tropical storms were 'seeded' and made to yield their rain before they had a chance to develop into destructive hurricanes (which do millions of pounds' worth of damage each year around the Gulf of Mexico), thus upsetting an important mechanism transferring tropical warmth to the middle latitudes. Hurricane modification could make our winters colder and increase our heating bills, and thus lower our standard of living. Weather control and the resulting accurate long-range weather forecasting would create many other problems in every walk of life. If it were known that the weather would definitely be wet and cool on the south coast and hot in Lancashire, then a crush of people would try to have their holidays on the favoured coast. It is perhaps after all in the interests of us all that the Meteorological Office should continue to produce many forecasts which, to the best of their knowledge, will prove correct but which, due to natural causes, will actually prove wrong.

NOTES TO CHAPTER 24

1. M.O. Bracknell, *Climatological Memorandum,* No. 73, Table 2.
2. B. Knapp and S. Child, 'Hydrological effects of man's activities', *Teaching Geography*, 5, 1979.
3. N. Calder, *The Weather Machine*, BBC, 1974.
4. Virginia Woolf, *Orlando*, Bloomsbury, 1993.
5. G. Manley 'Central England temperatures: monthly means 1659–1973', Q.J. 100, pp. 389–405.
6. J.G. Ballard, *The Drought* (novel), Penguin, 1968.
7. A. Barry, 'Nothing like a nice coal fire to set Noah off again', *Daily Telegraph*, 10 October 1977. 'Carbon-dioxide increases bring temperature rise', *The Times*, 17 October 1977. Nature-Times News Service: 'Rise in sea-level could drown coast areas', May 1975; 'Ice ages: studying the earth's orbit', November 1974; 'Climate: ice cores and the Vikings', May 1975. A. Tucker, 'Air of uncertainty—the ozone layer', *Guardian*, 22 January 1975. H.H. Lamb, *Climate, Present, Past and Future*, Methuen, 1977.

Further Reading

A number of the books listed can only be found in libraries or second-hand book-shops.

J.G. Ballard, *The Drought* (novel), Penguin, 1968.

D. Bowen, *Britain's Weather*, David & Charles, 1973.

J.H. Brazell, *London Weather*, HMSO, 1968.

W.J. Burroughs, *Watching The World's Weather*, CUP, 1991.

W.J. Burroughs, *Mountain Weather, A Guide for Skiers and Hill Walkers*, Crowood Press, 1995.

R. Bush, *Frost and the Fruitgrower*, Cassell, 1945.

N. Calder, *Spaceship Earth*, Penguin Books, 1991.

N. Calder, *The Weather Machine—and the Threat of Ice*, BBC, 1974.

C. Carter, *The Blizzard of '91*, David & Charles, 1971.

T.J. Chandler & S. Gregory, *The Climate of the British Isles*, Longman, 1976.

E. Cox, *The Great Drought of 1976*, Hutchinson, 1978.

I. Currie, *Red Sky at Night: Weather Sayings for All Seasons*, Frosted Earth, 1992.

I. Currie, *Frost Freezes and Fairs: Chronicles of the Frozen Thames and Harsh Winters in Britain from 1000 AD*, Frosted Earth, 1996.

I. Currie & M. Davison, *London's Hurricane*, Froglets, 1989.

I. Currie & M. Davison, *Surrey in the Hurricane*, Froglets, 1988.

I. Currie & M. Davison, *The Surrey Weather Book*, Frosted Earth, 1996.

I. Currie, M. Davison, & R. Ogley, *The Berkshire Weather Book*, Frosted Earth & Froglets, 1994.

I. Currie, M. Davison, & R. Ogley, *The Essex Weather Book*, Frosted Earth & Froglets, 1992.

I. Currie, M. Davison, & R. Ogley, *The Hampshire and Isle of Wight Weather Book*, Frosted Earth & Froglets, 1993.

I. Currie, M. Davison, & R. Ogley, *The Kent Weather Book*, Frosted Earth & Froglets, 1993.

I. Currie, M. Davison, & R. Ogley, *The Norfolk and Suffolk Weather Book*, Frosted Earth & Froglets, 1994.

I. Currie, M. Davison, & R. Ogley, *The Sussex Weather Book*, Frosted Earth & Froglets, 1995.

P. Damari, *The Herefordshire and Worcestershire Weather Book*, Countryside Books, 1995.

E.R. Delderfield, *The Lynmouth Flood Disaster*, E.R.D. Publications, Exmouth, 1976.

D. File, *Weather Facts*, OUP, 1991.

A. Glenn, *Weather Patterns of East Anglia*, Terence Dalton, Lavenham, 1987.

J. Gribbin, *Weather Force*, Bison Books, London, 1979.

G. Hill, *Hurricane Force*, Collins, 1988.

I. Holford, *Guinness Book of Weather Facts and Feats*, Guinness Superlatives, 1982.

I. Holford, *British Weather Disasters*, David & Charles, 1976.

B. Horton, *West Country Weather Book*, published by Barry Horton, 4, Hill Street, Totterdown, Bristol BS3 4TP, 1995.

J. Hurley, *Snow and Storm on Exmoor*, Microstudy E1, Exmoor Press, 1972.

R. Inwards, *Weather Lore*, S.R. Publishers, Wakefield, 1972.

J.L. Kerr, *The Great Storm*, Harrap, 1953 (the east-coast floods).

Professor H. Lamb, *Climate History and the Modern World*, Methuen, 1982.

D.I.C. Ludlam & R.S. Scorer, *Cloud Study, A Pictorial Guide,* John Murray, 1957.

Professor G. Manley, *Climate and the British Scene*, Collins-Fontana, 1970.

Meteorological Office, *Observer's Handbook*, 4th edition, HMSO, 1982.

Meteorological Office, *Meteorological Glossary*, 6th edition, HMSO, 1991.

M.D. Newson, *Flooding and Flood Hazard in the UK*, OUP, 1975.

R. Ogley, *In the Wake of the Hurricane, National Edition*, Froglets, 1988 (16 October, 1987).

M. Pollard, *North Sea Surge: Story of East Coast Floods of 1953*, T. Dalton, 1978.

D.N. Robinson, *The Louth Flood, 1920*, The Museum, 4 Broadbank, Louth, Lincs, 1995.

Sir N. Shaw, *The Drama of Weather*, CUP, 1940.

L.P. Smith, *Seasonable Weather*, Allen & Unwin, 1968.

D. Summers, *The East Coast Floods*, David & Charles, 1978.

A. Tinn, *This Weather of Ours*, Allen & Unwin, 1949.

M.J. Tooley & G.M. Sheail (editors), *The Climatic Scene*, Allen & Unwin, 1985.

A. Watts, *Wind and Sailing Boats*, David & Charles, 1973.

J.H. Willis, *Weatherwise*, Allen & Unwin, 1944 (Norwich weather).

N. Winkless & I. Browning, *Climate and the Affairs of Man*, Davies, London, 1976.

Some Suppliers of Meteorological Instruments

AGI-Obsermet Ltd., Ebblake Industrial Estate, Verwood, Dorset BH13 6BE

Campbell Scientific Ltd., 14–20 Field Street, Shepshed, Leics LE12 9AL

Casella London Ltd., Regent House, Wolseley Road, Kempston, Bedford MK42 7JY

Darton & Co. Ltd., Mercury House, Vale Road, Bushey, Watford, Herts WD2 2HG

Delta-T Devices Ltd., 120 Low Road, Burwell, Cambridge CB5 0EJ

ELE International Ltd, Eastman Way, Hemel Hempstead, Herts HP2 7HB

ICS Electronics Ltd., Unit V, Rudford Industrial Estate, Ford, Arundel, West Sussex BN10 0BD

MET-CHECK, P.O. Box 284, Milton Keynes, MK17 0QD.

R & D Electronics, Beaufort House, Percy Avenue, Kingsgate, Broadstairs, Kent CT10 3LB

Ultra-Pro, Unit 34, Wharf Street, Warwick, CV34 5LB

Unidata Europe Ltd., Enterprise House, Lloyd Street North, Manchester, M15 4EN

Vector Instruments, 115 Marsh Road, Rhyl, Clwyd LL18 2AB

W.J. Read, 49 Old Vicarage Park, Narborough, King's Lynn, Norfolk PE32 1TH

General Index

Note: Entries for places are given when they are mentioned in specific weather records or events, or tables of data. General surveys and tables of record weather will also be found under, for example, 'rainfall, extremes', or 'drought (1976)'. Figures in italics refer to illustration captions or to tables. Dates indicate that the reference is to a specific event. Numbers on the left indicate location on the map on page 294 (overleaf). Abbreviations used: av. = average, temp.= temperature, varn. = variation.

Shetland
Isles

1

2

3

4 ● Fair Isle

Outer Hebrides

Stornoway
11

13
14
12

15

16 17

10
9 8

5
6 Orkney Isles
7

4 ● Fair Isle

111
112 ● Carlisle
113
Penrith
Cross Fell
110 △ 108
109 △
114
LAKE DISTRICT
Keswick
129
128 △ 131
127 △ Ullswater
126 △ 124
130 △
Derwentwater
132 △
123
Ambleside
122
121 120 Windermere
Coniston 119 Kendal
Water
117 ● Ingleton
10 miles
16 km
118 ● Morecambe

113
Appleby
115
Shap
116

Solway Firth

19
18
33
32 39 40
23
21 22
31
26 28 29 27 30
24
25
59

34
35 36 37
38
42 43
44 45 46
52
41
51
55 54
57 71 72
56
70
69 67 68 76
66 65 74 77 75
Glasgow 78
63 88 86 85
64 89
47 Aberdeen
48
49
50
53
Bell Rock
73
Edinburgh 79 80 81
84 83
82
92
91 △93
90
62
61
60

NORTH CHANNEL
553
554 555
556 557 560
558 559
Belfast
561

Isle of Man
549
552
550
551

553

NORTH
SEA

430 431
432
433
434 △
435 436 437
438
442 444
446
457 455 454
456
458 480
497 481
496 482
495 488
494 493 485
501 498 500 499
502 503 511 506 510
507 508 509 512
504 505
Thames
440 441
443
424 425 426
423 419
448 447 417 414
449 451 420 418 471
450 452 461 463 462
459 460 472
479 491
483 477 475
484 478 473
489 485 △476
492 490
515
513
514
427
412
416 413
466
465 469
468
470 535
519
518 520
521 522
516 517
511
525
526

E I R E

Dublin
562

563

564
566
565
567

571

568
569

570

FASTNET

111
LAKE
Seathwaite
128
DISTRICT
123 119
118
Blackpool
Anglesey
241
240 239
236 235
242 244
243 △
245 247
246
251 250
249
254
255
258
259 260 265
263 264
261 262 266
Swansea
268 270
303 302
306 305 304
308 307
309
322 321
333 332
311 310
312
316
313 314 315

90
95
96 97
98 99
100
102
103
105 104 136
106 107 135 138 141
108 134 137 140 139 Scarborough
142 143
133 145 144
149 147 146
150 148
151 155 156 170 171 172
153 154 159 157
161 160 173 176
163 164 165 166 169 174 175
232 Manchester 168
231 233 162 167 189
229 230 222 193 194 195
228 227 220 192 196 198
226 224 223 218 196 197 203
225 219 216 205 207
223 217 214 215 206
256 289 290
257 286 287 291 292
276 △ 277 284
274 275 285 284
269 272 278 283
267 282 293 294
281 296 368 369
279 280 298 367 371
300 299 360 362 361
301 341 342
336 334 343 350 351
337 340 344 348 349
338 345 347 346
Cardiff Bristol
Birmingham
Liverpool
94 Newcastle
Hull 177
178 181
180 179
182 183 184
185
186
187
188 200
201 204
202
390 391
388 389 392 393 399 396
387 386 394 395 397 398 400
383 384 385
381 382
380 377 378 379
401
372 373 412 411 405 403 402
374 375 410 409 406 404 539 540
457 450 Kew 408 407 541
478 461 536 544 545
516 Wisley 534 542 543 533
Gatwick 523 530 531
524 528 529
507 503 511 512 511
355 356 527
353
359 354
Brighton
London
Heathrow
Cambridge

Bristol Channel
Penzance
Scilly
Plymouth
ENGLISH
CHANNEL
Bournemouth

Alderney
546
Guernsey
547
Jersey
548

80 miles
50 km

Index of Authors